Philosophy and Spacetime Physics

Philosophy and Spacetime Physics

LAWRENCE SKLAR

University of California Press   Berkeley   Los Angeles   London

University of California Press
Berkeley and Los Angeles, California
University of California Press, Ltd.
London, England
© 1985 by The Regents of the University of California
Printed in the United States of America
1 2 3 4 5 6 7 8 9

Library of Congress Cataloging in Publication Data

Sklar, Lawrence.
  Philosophy and spacetime physics.
  Includes index.
  1. Space and time. 1. Philosophy and science.
I. Title.
QC173.59.S65S54 1985   530.1'1   84-24128
ISBN 0-520-05374-5

To Jessica

Contents

Acknowledgments

Research contributing toward the papers in this volume has received support from the John Simon Guggenheim Memorial Foundation, the Rackham Foundation, and the National Science Foundation. The generous help of these foundations is gratefully acknowledged.

I also wish to acknowledge the help of the original publishers of these papers, whose permission to reprint them here has made the appearance of this volume possible.

"Methodological Conservatism." *Philosophical Review* 84 (1975): 374–400. Chapter 1.

"Saving the Noumena." *Philosophical Topics* 13 (1982): 89–110. Chapter 2.

"Facts, Conventions, and Assumptions in the Theory of Spacetime," in J. Earman, C. Glymour and J. Stachel (eds.), *Foundations of Space-Time Theories. Minnesota Studies in the Philosophy of Science,* vol. 8. Minneapolis: University of Minnesota Press, 1977, 206–274. Chapter 3.

"Do Unborn Hypotheses Have Rights?" *Pacific Philosophical Quarterly* 62 (1981): 17–29. Chapter 4.

"Modestly Radical Empiricism," in P. Achinstein and O. Hannaway (eds.), *Observation, Experiment, and Hypothesis in Modern Physical Science.* Cambridge: MIT Press, 1985, 1–20. Chapter 5.

"Inertia, Gravitation and Metaphysics." *Philosophy of Science* 43 (1976): 1–23. Chapter 6.

"Semantic Analogy." *Philosophical Studies* 38 (1980): 217–234. Chapter 7.

"Incongruous Counterparts, Intrinsic Features, and the Substantiviality of Space." *Journal of Philosophy* 71 (1974): 277–290. Chapter 8.

"What Might Be Right about the Causal Theory of Time." *Synthese* 35 (1977): 155–171. Chapter 9.

"Prospects for a Causal Theory of Space-Time," in R. Swinburne (ed.), *Space, Time, and Causality,* Dordrecht: Reidel, 1983, 45 – 62. Chapter 10.

"Time, Reality, and Relativity," in R. Healy (ed.), *Reduction, Time, and Reality.* Cambridge: Cambridge University Press, 1981, 129 – 142. Chapter 11.

"Up and Down, Left and Right, Past and Future." *Nous* 15 (1981): 111 – 129. Chapter 12.

Introduction

Three major themes dominate the issues treated in the essays collected in this volume:

First, given the plausibility of the allegation that our geometric theories of the world outrun in their content the possibility of unique specification by all possible empirical data, how can we possibly avoid the skeptic's assertion that we ought not to claim to have genuine knowledge of the geometry of the world at all? Further, doesn't the skeptical threat present in the geometric case generalize to a threat of skepticism against alleged theoretical knowledge in general?

Second, ought we to view the apparent ontology of geometric theories, space and time (or, rather in the relativistic context, spacetime), as genuine constituents of the world, or ought we to, instead, interpret the apparent theoretical ontology of geometry in some instrumentalistic fashion, taking the reference only of the observational terms of the theory (light rays, particles, measuring rods and clocks, etc.) straightforwardly? Once again the ontological question asked here generalizes from the geometric to the general theoretical context.

Third, and finally, how ought we to evaluate those philosophical doctrines which claim a reducibility, in some important sense, of the concepts of geometric theory to concepts not prima facie geometrical at all? In particular how plausible are the claims that metric or topological aspects of spacetime are reducible to causal relations among events, and how plausible are the claims that the temporal asymmetry of the world is reducible to the asymmetry in time of features of the world captured by the application of the concept of entropy?

Naturally these questions are not totally isolable from one another, for any answer to one of the questions posed above is replete with consequences for one's position regarding answers to questions in the other categories. For this reason many of the essays included here treat of problems in all three categories, although it might be plausible to claim that the emphasis in the first five chapters is on epistemic matters, of the next three, on questions of ontology, and of the last four, on issues of causal or entropic reducibility of spacetime concepts.

The essays in this volume have all appeared subsequent to my earlier book on the philosophy of space and time, *Space, Time, and Spacetime*. While the reader of the present volume who is unfamiliar with the background of contemporary philosophical discussion of issues concerning space and time would find that book helpful in setting the stage for the discussions contained in these essays, the present introduction, along with the prefaces to the chapters in this volume, should make the context of argument sufficiently clear so that the reader can grasp the motivation for the arguments contained here without an extensive earlier background in the philosophy of space and time. For those familiar with the earlier work, however, it might be helpful to note several ways in which these essays hope to go beyond it.

In some cases, recent developments in the foundational study of spacetime theories in physics has opened up new areas for philosophical inquiry and allowed for a more profound treatment of older issues. For example, the intensive study of the extent to which variously construed causal structures do or do not uniquely determine spacetime structures in models for general relativity has greatly expanded our opportunity to clarify aspects of causal theories of spacetime in philosophy. In other cases the rediscovery of earlier but neglected parts of foundational physics has been of importance. This is the case, again of importance when one is dealing with causal theories of spacetime, with the recent attention paid to Robb's early but frequently overlooked formulation of special relativity. Finally, there were some important aspects of the scientific framework which I felt simply had not received the attention they deserved, given their philosophical relevance, in the earlier work. Trautman's curved spacetime rendition of Newtonian gravitational theory, for example, failed to receive the attention it merited in *Space, Time, and Spacetime*.

From the philosophical side, recent years have seen a diverse, yet related, number of attempts to reconcile a realist attitude toward theories with an antiskeptical epistemology, while at the same time acknowledging the force of the empiricists allegation of underdetermination of theory by the totality of possible empirical data. Such approaches offer an account of theoretical equivalence more stringent than that of the positivist, who dubs all observationally equivalent theories fully equivalent; accounts of confirmation which attempt to show us that even if they are fully observationally equivalent, two nonequivalent theories may be differentiated in their worthiness for our belief; and accounts of meaning, of reference, and of explanation which, again differing from those of the positivist, attempt to show us why we ought to take the putative ontologies of theories at realistic face value and not construe them away in some eliminationist or instrumentalistic manner.

Many of the essays in this volume are directed at such realist accounts of theory, not primarily to decide whether or not they are correct, but, rather, to differentiate alternative realist approaches from one another in important ways, to lay out in as full a way as possible the necessary presuppositions which must be made to make the various approaches plausible, and, finally, to make it as clear as I can just what the more traditional empiricist would claim to be the places where such approaches to theory are in serious want of further explication and justification.

Finally, on a few matters — for example, the entropic account of time asymmetry — there were substantial doubts on my own part that the treatment given in the earlier book did full justice to the philosophical issues involved. On this particular topic, some preliminary moves are made here to move in the direction of a fuller and fairer discussion of the issues involved, although much more remains to be done.

In the remainder of this introductory essay, I will in turn take up the three major areas of the epistemology of geometry, the ontology of geometric theories, and the nature of causal theories of space-time structure in the hopes of providing for the reader a brief survey of the background discussion in both physics and philosophy which provides the context in which the detailed arguments of the various papers find their place. Again, it should be noted that each essay will be preceded by its own short preface to fix its place in the general scheme even more closely.

The Epistemology of Geometry

For two millennia geometry provided philosophers with the paradigm of a science whose propositions could, allegedly, be known to
be true with certainty and independently of reliance upon observation and experiment. The discovery of the axiomatic non-Euclid-
ean geometries, and of the far more general n-dimensional generalizations of Gauss's theory of surfaces propounded by Riemann,
cast grave doubts on this traditional view of geometry as an a priori
discipline. The obvious direction in which to go was to suggest that
geometry was just one more theory of the world whose epistemic
basis rested on inductive generalization from the data of experience.

While there is plainly something correct about this empiricist
view, the need for a more subtle approach was emphasized by
Poincaré's demonstration that, were one willing to make sufficient
modifications elsewhere in one's physical theory of the world, any
geometry could be held as undisconfirmed by any body of observational data whatever. Most of the interesting work on the epistemology of geometry since his time has been in the varying responses to this alleged underdetermination of geometric theory by
all possible observational evidence.

One important response is that of the positivist. Offering an
account of theory as being nothing but a summary of lawlike regu-
larity among the observables, he counters the threat of skepticism
implicit in the underdetermination argument by adopting a notion
of theoretical equivalence which is such that any two theory expressions which give rise to the same observational consequences
are taken as merely formalistic variants, each expressing one and
the same underlying theory. If this is so, then there is no longer an
epistemic puzzle regarding alternative theories saving the same
phenomena, since there are no such genuine alternatives.

But, taken seriously and rigorously, this positivist solution has
consequences too irrealist for many to swallow. In the geometric
case it seems that we must dispense with spacetime itself as an
entity in the world, since equivalent (positivistically) to any spacetime theory will be the mere set of its observational consequences,
and if these refer only to material objects and their relations, then
since this new expression of the theory is equivalent to the old, we
ought not to have taken the theoretical ontology of the older ex-

pression at face value at all. Worse yet, if, as is frequently the case in geometric underdetermination arguments, we restrict the observable to the local relations among material measuring instruments, then, even relationistically construed, the global geometry of the world becomes a mere fiction.

One way to avoid these consequences would be to deny the common presupposition of the underdetermination argument and of the positivist reply to it. Both presuppose that it makes coherent sense to distinguish among the totality of consequences of a theory its proper subset of observational consequences. But many arguments have been advanced to try and convince us that such a notion of an in-principle delimitation of the realm of the observable is an empiricist mistake. Of course if we accept such a claim the problem as posed by Poincaré cannot be coherently framed.

But, at least in the geometric case, we have reason to be somewhat suspicious of this "way out." Far too many of the accomplishments of the best physical science which deals with the geometry of the world themselves rest upon an implicit assumption of an observational/nonobservational distinction, or, rather, one at least strong enough to assure us that some of the consequences of some of our theories must be considered in-principle nonobservable. Einstein's denial of the aether frame in the theory of special relativity, and his denial of the discriminability of a gravitational field from the metric of spacetime in general relativity, both rest, I believe, on such an assumption. So it certainly behooves us to take the possibility of such a distinction seriously enough to explore what alternatives are available to us assuming it to be a legitimate one.

One approach to the problem avoids the threat of skepticism by adopting some form or other of a permissive notion of rationality. Here it is argued that in the case of theoretical underdetermination we are rational to believe any one of the empirically well-confirmed theories. Which one? Any one we choose. In sophisticated versions such a notion of rationality may be combined with an elaborated pragmatist critique of the very notion of truth itself, and with a denial of the legitimacy of doubt except as internal to one's adopted way of "getting on in the world." But to some, myself included, such an approach seems to be merely skepticism sugar-coated so as to make its consequences appear more palatable.

It is worthwhile, in any case, to see just how far one can get in trying to simultaneously maintain a realist notion of truth, an antipositivist and antireductionist approach to theory, and an acceptance of the claim that our body of potential empirical evidence is outrun by the full contents of theory. Much contemporary "realist" work on the epistemology of space and time tries to do just that, and a good deal of the epistemological content of the essays in this volume is devoted to explicating ways in which this approach might go, and to pointing out pitfalls in the path of the philosopher trying to solve the problem in this way.

All such realist views will deny the fundamental positivist thesis that observational equivalence is a sufficient condition for full equivalence of theories. All will demand the possibility that two theories, although inequivalent, can equally well save all possible phenomena. What condition they will impose, over and above observational equivalence, for full equivalence will depend on the attitude they take toward the meaning of the nonobservational terms of the theories.

One view will argue that the nonobservational terms of a theory cannot have their meaning fully accounted for solely in terms of the role the terms play in the theory. One such approach, for example, will argue that terms which appear in observational contexts can appear, meaning exactly the same thing, in nonobservational contexts as well. Thus, it will be said, we mean the same thing by 'length' whether we are speaking of the length of a table or the length of the interatomic bond in a molecule. Such a "semantic analogy" approach to the meaning of terms designating nonobservables, or indeed any other approach which takes meaning attribution to outrun that which can be captured by examining the role played in the theory itself by the theoretical term, runs into many difficulties, both in offering a coherent account of meaning accrual for such terms and in offering a systematic account of what it takes, over and above observational equivalence, to make two theory expressions genuinely fully equivalent.

The alternative approach, which attempts to reconcile realism with a "role played in theory" approach to the meaning of the nonobservational terms of a theory, is quite a bit more theoretically tractable, and it is an approach which has appealed to many in recent years. This approach is frequently associated with a favorable attitude toward Ramsey's claim that we should properly view theories as adequately represented by the result obtained by (1) conjoining their axioms into a single assertion, and (2) replacing the nonobservational terms with existentially quantified predicate variables. From this point of view "charm" in physics simply is that property which the theory says exists, and which has whatever features the theory says charm has. Relations of charm to observable features are captured by the terms referring to these features staying intact in the Ramsey sentence, but relations to other nonobservable features are captured by the appearance in the Ramsey sentence of other existentially quantified predicate variables.

Common to this approach to theories is the assertion that what is required for full equivalence, over and above observational equivalence, is that the theories in question bear some appropriate structural isomorphism to one another at the nonobservational level. Theories which can be transformed into one another by term-by-term interdefinitions of theoretical terms, for example, we ought to hold genuinely equivalent, but theories whose theoretical structures are not sufficiently structurally similar ought not to be taken as fully equivalent even if they are observationally equivalent.

Such an approach to theories is well worth pursuing in detail, for, if successful, it might very well do justice to the legitimate epistemic and semantic claims of the positivist without being forced into his too generous notion of theoretical equivalence with its irrealist consequences. But there are many problems, both of detail and of principle, in such an approach, and much of the epistemological material in the essays in this volume is devoted to following some of these problems up. Most crucial to such an approach is the question of how it is to handle the skepticism latent in any view which lets theoretical content outrun empirical determinability.

A number of suggestions have been made which propose to solve this epistemic problem for the realist by invoking grounds for theory choice which outrun conformity with even all possible empirical data. Some invoke various notions of a priori plausibility for theories, taking, for example, ontologically simpler theories to be more worthy of belief. Here problems of characterization and of justification arise, as well as problems engendered by the plausibility of such claims as that the simplest theory is, ultimately, the positivist reduction of the theory which does away with nonobservational ontology altogether. Other moves focus on the alleged dependence of our principles for theory acceptance on the best available antecedent theory we have accepted to date. In this context such "reliabilist" views about justification of theory choice frequently cash-out as the view that one ought to believe the theory which saves the phenomena and is the most conservative change from our previously accepted theory. Here again serious problems arise, for it is not at all clear that such a notion will serve to coherently or consistently make a theory choice possible, nor that having done so an adequate rationalization of such a rule for choice from the realist perspective is possible.

Finally, there are those who seek the resolution of the realist's epistemic problems in the characterization of models of confirmation theory. Such, for example, is Glymour's "bootstrap" model of confirmation (expounded in his *Theory and Evidence*) which offers an account of confirmation in which the alternatives to the standard relativistic spacetimes receive no confirmation, and Friedman's invocation of notions of consiliance of inductions (in, for example, his *Foundations of Space-Time Theories*), which tries to explain why we ought, if we are seeking for highest confirmation, to understand the Ramsey sentence version of a theory physically-realistically instead of merely as an instrumentalistic device whose existential quantifications are only indicative of abstract representations.

Again these realist approaches to obviating the threat of skepticism are fraught with problems both of detail and of principle, and some of the critical possibilities are explored here in those chapters and portions of chapters which focus on the epistemological status of spacetime theories. Of course many open questions remain. Some of these are problems in the epistemology of theories in general. Other problems remain centering around the question of the degree to which spacetime theories present special and idiosyncratic epistemological problems which differentiate them from the ordinary run of naturalistic theories.

The Ontology of Spacetime Theories
and Their Explanatory Role

A persistent controversy rages in the philosophy of space and time between those who claim that spacetime ought to be taken as an entity which exists in its own right, and which could exist even were there no ordinary matter in the universe, and those who maintain that talk about spacetime ought to be considered as nothing but a misleading way of representing the fact that there is ordinary matter and that there are spatiotemporal relations among material happenings. Call the former doctrine substantivalism and the latter relationism.

Early philosophical debate on this issue frequently hinged on the substantivalist's claim that the relationist was unable to account for the legitimacy of all reasonably assertible propositions about spacetime — for example, about empty regions of it or what it would be like even if totally empty of matter; and the relationist claim that the substantivalist allowed for too prolific a language in which the illegitimate could be expressed — for example, that in substantivalist terms it made sense to ask where in space itself the material universe was located. Much of this debate can be found in the famous correspondence between Leibniz and the Newtonian Clarke.

But along with some of the best traditional philosophical argumentation on this issue which can be found there, especially in the form of Leibniz's ingenious formulation of relationism and his brilliant marshalling of the most effective antisubstantivalist arguments, we also find Newton's surprising attempt to demonstrate the truth of substantivalism from the experimental observation of the existence of the so-called inertial forces. According to Newton the distinction between real acceleration, relative to space itself, and mere relative acceleration, of one material object with respect to another, is one demonstrable by the observable causal effects of the real change of motion.

While the early relationist replies to Newton were uniformly ineffective, the centuries following Newton's seminal work saw numerous, quite varied, attempts to make the issues clearer. Some were attempts to make Newton's substantivalism more respectable by showing that if nonmechanical phenomena were taken into account features with no empirical consequence on the original view could be found to have observational consequences. For example, the aether theories of light could be taken to show, among other things, that uniform motion with respect to space itself, as well as accelerated motion, had empirical consequences.

Other approaches — not actually forthcoming until imaginations had been spurred by the invention of spacetimes appropriate for the new non-Newtonian theories of special and general relativity — were designed to show that we could construct spacetimes more appropriate to Newton's data than was Newton's absolute space. Neo-Newtonian spacetime is one which gives rise to the same empirical results as does Newton's space-through-time, but which is relationistically more acceptable, as within it such notions as absolute place through time and absolute velocity cannot be defined. A curved version of such a spacetime provides an arena for Newtonian gravitation again more relationistically acceptable than Newton's original spacetime. Yet both theories are, of course, substantivalist theories in the sense of taking spacetime to be an entity over and above the material inhabitants of the spacetime.

Finally, some approaches tried to reconcile the Newtonian observational data with a genuinely relationist theory. Mach, for example, tried to account for the inertial forces as forces arising from the acceleration of test systems not with respect to space itself, but, rather, with respect to the remaining averaged-out matter of the universe. Alternatively, one could offer a relationistically acceptable account by simply denying that absolute acceleration exists or that an explanation of inertial forces is called for. In such an account we can explain differences in inertial forces by relative motions of material objects, but we simply offer no explanation at all as to why some objects (the inertial ones) suffer no such forces at all.

With the discovery of special and general relativity the discussion must shift somewhat. Now it is the question of the appropriateness of substantivalist or relationist ontologies for spacetime to these new accounts of the world which is at issue, rather than to the discredited Newtonian account. While some early writers thought that special relativity was the long-awaited appropriately relationist theory, further reflection showed that in this account of the world the distinction between inertial and noninertial motion, with which the Newtonian underpinned his substantivalism, is just as strong (rather, stronger) in this account than in the Newtonian. And while Einstein was originally motivated to discover general relativity at least in part by the hope that he was coming up with a theory which was relationist in the Machian vein, further thought showed that this was not a very plausible reading of what the theory implied.

Indeed, on a surface reading, the general theory of relativity certainly has the appearance of a substantivalist account. Of course there is no Newtonian space-itself through time, nor Newtonian absolute position or velocity. Nor are there the global inertial frames of Newtonian, neo-Newtonian or Minkowski spacetime. But there is the geodesic structure of a curved spacetime, a structure whose observable effects on the motion of material objects is plain. And there is the variable metric of this curved spacetime which, on at least a surface reading of the theory, exists whether or not measuring rods and clocks are there to determine it. Again, on general relativistic grounds the notion of spacetime itself, even if devoid entirely of ordinary matter, with its own structural features seems to make perfectly coherent sense and sounds substantivalist indeed.

But we cannot be satisfied with such a surface reading of the metaphysics of the theory. We must, instead, probe more deeply into the philosophical sources of the arguments relied on by substantivalists and relationists. Much of the impetus behind the relationist attack on substantivalism is, of course, motivated by that same skepticism regarding the unobservable which constituted the basis of the epistemological debates about geometry. Relationism is plainly a doctrine akin to other "instrumentalistic" doctrines regarding theories, doctrines which attribute genuine reference only to the names and predicates of the theory which aim to denote observable entities and properties, and which treat the apparently denoting terms which allegedly refer to nonobservable entities and properties as not really referring in nature at all. Instead, the linguistic apparatus of a theory which superficially refers to the unobservable is taken to function rather as an intermediary device allowing us to move inferentially from observation to observation, but not as making genuine existential claims of its own at all. Much of the argumentation here is familiar from the positivist epistemic critique of theoretical realism discussed earlier.

But the relationist argument against substantivalism has curious idiosyncratic elements of its own, elements which differentiate it from the usual epistemically motivated attacks against allegedly otiose ontology. These special arguments rest upon various symmetry considerations concerning space and time. Essentially, the arguments go, since the points of space (or spacetime) are everywhere alike and since all directions in space (or spacetime) are alike, that is, because space (or spacetime) is, allegedly, homogeneous and isotropic, the substantivalist view commits us to distinctions between possible worlds which are not real distinctions, and is, hence, an illegitimate doctrine. Consider, for example, Leibniz's famous arguments to the effect that if substantival space existed, there would be a difference between the actual world and a world which differed from it only in the place of the material universe in space itself. But such a difference would not be a genuine qualitative difference. Hence, by a version of the principle of the identity of indiscernibles, the two possible worlds would have to be the same possible world, refuting the substantivalist's claim that they are genuine alternatives.

Some characteristic objections of the substantivalist to the relationist, and of the relationist to the substantivalist, come as no surprise. Crudely, many substantivalists, besides supporting their doctrine on the kinds of positive grounds which are the modernizations of Newton's important argument, criticize relationism for its tendency to lead us to an extreme irrealism, the result of following out the relationists' arguments against substantival spacetime to, by parity of reasoning, arguments against a world of independent (of mind) existence at all. On the other hand, the relationist will frequently argue that one is forced to the full-fledged relationist denial of substantival spacetime by exactly the same sort of reasoning which, within natural science, leads us to deny the existence of otiose theoretical postulates — for example, the denial of the aether frame in special relativity or the denial of global inertial frames in the general theory of relativity. The relationist will frequently claim that it is only a failure of nerve on the part of the substantivalist which allows him not to see that exactly the same sorts of thinking which leads us to eschew the aether frame ought to lead us, by consistency of reasoning, to eschew substantival spacetime altogether.

It is at this point in the argument that the differential attitude toward scientific explanation of the substantivalist and the relationist frequently becomes crucial. The substantivalist will often argue that the need for a residual substantivalist spacetime in our conceptual theory is the need for the postulation of features of the world which *explain* the observational data. Just as Newton requires space itself to explain the inertial forces observed, so in general relativity we need the geodesic structure and metric of spacetime itself to explain the mechanical, optical, and metrical facts available to us in our empirical data. The relationist counter to this is to critique the substantivalist's notion of explanation in several important ways. A careful examination of the interrelation of epistemic, semantic, and ontological views about theories with models of just what constitutes an explanation in science is needed here. For it is no trivial matter to resolve the debate between the substantivalist, who tells us that the relationist can only summarize the observational data and not explain it, and the relationist, who tells us that the substantivalist's alleged explanatory apparatus is only pseudoexplanatory and a redundant inflation of our reasonable scientifically established ontology of the world.

While most modern substantivalists argue for their doctrine as being necessary in the light of the ontological commitments of our best available spacetime theories, in particular as implied by the general theory of relativity, there are other characteristic substantivalist arguments as well. One version of substantivalism, for example, argues to the necessity of this ontologically realist doctrine from the existence of global topological facts, such as the orientability (or nonorientability) of the spacetime. Here the style of the argument traces back in its historical origins to Kant's well-known, but enigmatic, remarks about handedness and space. Once again the relationist has a body of counterarguments to this "outside the mainstream" argument for the existence of spacetime itself as an entity over and above the ordinary material constituents of the world.

There are special features of the substantivalist-relationist debate which arise out of the specific nature of our current best spacetime theory, the general theory of relativity. While some of these issues are touched on in the essays in this volume, I believe that profitable further exploration of them is called for and can be expected in the future.

One issue arises because the implicit distinction between ordinary matter and spacetime itself, a distinction simply assumed in the classical substantivalist-relationist debate of earlier years, is a distinction it is hard to maintain coherently in the face of general relativity. The identification of gravity with curvature of spacetime itself in the general theory of relativity leads to the conclusion that spacetime itself can embody mass-energy (although the lack of symmetry in the general spacetime model of the theory leads to grave questions of the global definability of the energy of the spacetime field). In the light of this, it simply isn't clear how the distinction between spacetime itself and "ordinary matter," a distinction so frequently utilized by the ordinary relationist, can be maintained. This problem for the relationist becomes even more crucial if one moves to some kind of "geometrodynamic" style of theory in which ordinary matter is *identified* with curved spacetime itself, rather than as an autonomous "inhabitant" of a spacetime arena. Further, traditional relationism relies quite heavily on Leibnizian arguments which rest on presupposed symmetries of spacetime. But in general relativity, where the models for the theories are, in general, nonsymmetric, these arguments lose much of their force. The older relationist arguments which tell us that, matter aside, each place in spacetime and each direction in it is just like every other are dubious indeed, in a theory in which intrinsically spacetime features, like the magnitude of curvature, can discriminate pure spacetime event locations from one another.

I think that both of these issues require the relationist to carefully study how his arguments against substantival spacetime fit into broader philosophical schemes. While not inevitable, a predictable relationist response to the first problem above will be to move ordinary matter into the same dispensable category as substantival spacetime, accepting the substantivalist's claim that a consistent relationism ultimately results in a phenomenalistic approach to theories in general.

Responding to the second problem noted above will require the relationist to think through issues concerning identity and diversity with greater thoroughness. Even without symmetry, he will argue, the idea of spacetime itself is illegitimate. While structural features, indeed those with observable causal consequences, will, in a nonsymmetric spacetime world, differentiate event-locations from one another by their purely spacetime features alone, the possibility of saving the phenomena by moving to an alternative world where the structure is the same but the identity of the spacetime locations having the structure is changed by a permutation of locations through structure will, he argues, show the illegitimacy of the substantivalist viewpoint. Here the consistent relationist will be led to consider the claim that just such a change, which keeps structure the same but which changes the identity of entity suffering the structure, is possible in the world of material things as well. And, once again, the consistent relationist will probably acknowledge that his antisubstantivalism vis-à-vis spacetime must, to be coherent, be generalized to an antirealism regarding ordinary matter as well.

At this point the relationist will need also to carefully examine his notion of identity and diversity, and, given his bent toward critiquing semantics from an epistemic point of view, probably look toward our basic notion of self-identity as the phenomenal origin of the notions of identity as projected onto the world of matter and of spacetime itself by the substantivalist.

On both these issues, taking the debate about spacetime as they do into some of the most intractable of general philosophical debates, we need far more clarity and enlightenment than is available to date, and, I believe, further work on these problems in the general philosophical context, and in the context of the specific issues governing the substantivalist-relationist debate about spacetime, will interact in ways which will be mutually clarificatory.

Causal Order and Spatiotemporal Order

A persistently recurring theme in the philosophy of space and time is that there is an intimate connection between the spatiotemporal order of events in the world and the causal order among these events. Sometimes it is alleged that the connection is so fundamental that we ought to view the spatiotemporal structure of the world not as a primitive feature of it, but, rather, as a feature which ought, in some sense, to be "defined away" in terms of, or "reduced to," the causal order.

The thesis is frequently maintained in an ontological vein, for example, with claims to the effect that spatiotemporal structure is "nothing but" causal structure. Or that we ought to take as the genuine, or real, spatiotemporal structure of the world only that spatiotemporal structure which can, in some appropriate sense, be identified with causal structure. Frequently the doctrine of the ultimately causal nature of spatiotemporal structure is also alleged to have important consequences of an epistemological sort as well, as in, for example, claims to the effect that insofar as spatiotemporal structure is not fully determined by some appropriate causal structure, the attribution of the spatiotemporal structure to the world ought to be taken as merely "conventional" in some important sense.

Before one can judge the plausibility of a causal theory of spatiotemporal order, however, one first must have a clear idea of what claims the alleged causal reducibility thesis is actually making. As a number of essays in this volume attempt to show, this is none too easy a thing to do. For, as these essays argue, there is a wide variety of quite different, and indeed sometimes incompatible, claims that different theses of causal reducibility are making. Sometimes the differences between these claims can be seen as forthright components of the doctrine. But in other cases various exponents of causal theses seem not to have been clear in their own minds about just which version of a causal theory of spacetime they want to expound and defend.

Sometimes causal theses are being offered as a kind of reduction of the spatiotemporal order to the causal which parallels that alleged in phenomenalism as a reduction of material-object language to the language of immediate content of perception. In theories of this kind it is usually claimed that our entire epistemic access into spacetime structure is through causal structure. It is for this reason that we ought to take assertions about spatiotemporal structure to reduce to assertions about causal structure in their very meaning.

Sometimes, rather, the claim is that spatiotemporal structure reduces to causal structure in the sense that the theory of matter reduces to that of atoms or the theory of light reduces to that of electromagnetism. In these cases the claim is that structure of one kind reduces to that of the other because it is a fact of nature, discovered by the progress of empirical science, that the entities discussed at one level of discourse are simply identical to those referred to by the other theory. So, it is sometimes claimed, tables are just arrays of atoms, light waves are just electromagnetic waves, and, in a similar vein, spatiotemporal relations are just causal relations.

Various causal theories of spacetime are also differentiated by what they take the appropriate causal relations to be which constitute the basis of the causal structure to which the spacetime structure is to be reduced. For some authors the basic causal notion is that of one event having a determining influence on the occurrence of another. For others it is the possibility, rather than the actuality, of determining influence which is crucial. Other accounts rely in a fundamental way on the fact that in a relativistic world causation is taken to be mediated by the connection of the events by a continuous causal signal. Here it might be connection or connectability which is the basic causal notion. Or, rather, it might be the structure of the causal propagations themselves which is essential. These distinctions may sound trivial, but they are not, and quite different causal theories of spacetime result, depending on which of these options one adopts in one's theory. Finally, for other authors it is the algebraic structure among quantum measurements which constitutes the real causal order to which spacetime order is to be reduced.

The kind of relation between spatiotemporal and causal feature which is essential for the correctness of the causal theory again varies from theory to theory. Need the causal and spatiotemporal relations only be coextensive for the "definition" or "reduction" to be counted as established? Or is it lawlike coextensiveness which is demanded? Perhaps some stronger kind of necessity for the coextensiveness is what really is in mind, say the kind of necessity which is the result of an identification of one relation with another. Or, even stronger, is it an analytic meaning reduction among the relations which is being called for? Once again, there is a wide variety of causal theories of spacetime, with differing presuppostions and differing intended claims. They must be carefully sorted out before we can even begin to examine their plausibility and their immediate or remote consequences.

Finally, just what aspects of spatiotemporal structure are supposed to be accounted for causally? In some theories it is the topology of spacetime which is reduced to a causal notion. In others it is the metric of spacetime which is also to be accounted for causally. On top of this there is the very special theory which tells us that it is the asymmetry or directionality of time which is to be reduced to a notion not prima facie spatiotemporal. Here the alleged reduction is usually taken to be to the entropic increase of systems in the world in time, a notion not properly thought of as causal at all!

Pursuing these issues sometimes requires the careful consideration of a variety of contemporary physical theories in some detail. One example comes from the philosophical examination of the special theory of relativity. It is frequently claimed that simultaneity for distant events in the special theory of relativity is merely a conventional notion. Sometimes it is alleged that this is not true in the prerelativistic spacetime context, and that the ground of the distinction rests upon the fact that simultaneity is causally definable prerelativistically but not in the special theory of relativity. But consideration of the early work of Robb on relativity shows that, in certain senses, simultaneity, and indeed the entire metric of spacetime, is causally definable in the relativistic context. Understanding just what Robb's work does and does not show about the inter-relation of causal and spatiotemporal structures in relativistic spacetimes; just what the relevance of his results is for various claims of definability or reducibility of spacetime to causal notions; and just what the relevance of these results to various allegations of conventionality is, is a subtle and difficult task. But it is worth pursuing, for in doing so one learns how many issues can be confounded together, and how much misdirection can result from practicing philosophy by slogans instead of by hard and detailed considerations of physics and philosophy.

Another fascinating and difficult test case is the study of the relation of topological structure to causal structure in the general theory of relativity. Here a full understanding of the physics requires that one bring to bear some of the most subtle considerations of the mathematics of spacetime structures. And the relevance of the results obtained to various philosophical questions is, once again, a matter which requires a good deal of thought in its own right. Ultimately, one discovers such results as the implausibility of causal theories of topology, construed in some ways, and the plausibility of other construals of causal theories of spacetime topology. But the plausibility of the latter versions is purchased, perhaps, at the price of realizing that what one called a causal theory really wasn't reasonably so designated at all.

Finally, on the alleged reducibility of the asymmetry of time to entropic features of the world, we again discover a host of difficult questions, many of them unresolved at the present time. While the essays in this volume take up some of these issues, in particular asking just what kind of reduction the alleged reduction is supposed to be, many difficult questions regarding the very nature of the concept of entropy, and of the structure of the statistical mechanical theory which defines it and in which it plays its explanatory role, remain to be given the close philosophical scrutiny they deserve.

The essays in this volume do pursue, at a certain level of abstraction, the general questions of types of alleged reducibility and definability, and some of the intrinsic difficulties of such claims, irrespective of the details which differentiate them, are brought to light. One issue which is hardly touched on here, and one which again requires far more philosophical attention than it has received, is the fact that almost all of these discussions on causal order and spatiotemporal order take place in a background which assumes the prequantum view of the world. Just what transformation the arguments must take when one accepts the necessity of viewing the whole of physics through the conceptual scheme of quantum theory remains to be seen.

Reflections on These Essays

The purpose of these essays is not to expound and defend some one particular approach to the philosophy of science, but, rather, to explore in some detail the presuppositions and consequences of several alternative points of view. All too often philosophical debate becomes a matter of sloganeering, with, for example, "realists" opposing "fictionalists" where it hasn't been made clear just what the opposing points of view really amount to. Semantic issues regarding the meaning of the terms which function in our theories, epistemological issues regarding the warrant for belief or disbelief we might have in theories, ontological issues regarding the degree to which we ought to take the apparently referential language of theories as genuinely referential, and such methodological issues as what is to count as explanatory in science, form a complex and tangled whole. Unless we go a good deal further than has been done in disentangling some of the relevant strands, I believe that further progress will be much impeded.

The method which I primarily apply to gain insight into some of these complexes of issues is to confront philosophical doctrine with contemporary physical theory. The viewpoint held here is most certainly *not* that one can resolve philosophical issues in a simple-minded way by reference to the results of our best available current scientific theory. Quite the contrary, the essays try to show that the work of theoretical science takes place in a context in which various philosophical presuppositions are, consciously or unconsciously, continuously being utilized to reach theoretical conclusions. The hope is that by seeing how we do behave in resolving open theoretical questions in science, some light will be thrown on the philosophers' questions of how we ought to proceed methodologically, and how such methodological stances ought to reflect underlying semantic, epistemological, and ontological consequences of philosophical standpoints.

Once again, the issues will be complex. First, one must realize that at the frontier of theoretical science there is no simple "what science tells us." Rather, what we find is a complex body of theory, including, on occasion, incompatible scientific alternatives between which theoretical science cannot decide. Even when we look back at older results in science we sometimes discover that a kind of "openness" is present which allows us to view the very results of scientific research from several, not obviously compatible, points of view. The myth of a monolithic uncontroversial body of scientific theory, contrasted with the squabbling indecision of the philosophers who never agree on anything, is a myth, although one not without the degree of truth most stereotypes possess.

The complexity and fluidity of the issues treated here suggest that a rather "dialectic" means of investigation is the appropriate one, and it is such a method that the reader will discover in these essays. Issues in philosophy are taken up over and over, each time from a somewhat different perspective. Single philosophical controversies are tested against a multiplicity of scientific results on which they bear, and single scientific theories are repeatedly examined from a multitude of philosophical perspectives. Again and again it is emphasized that philosophical issues are highly interdependent, and that one cannot opt for a solution to one problem without accepting the relevant consequences, sometimes unwelcome, which follow in other areas. And, again and again, the necessity for an open-minded attitude with regard to theories in physics is emphasized, with an insistence that we ought not to be satisfied with confronting philosophy by the standard theoretical view understood in the most common way, but, instead, must look at the wide variety of theoretical options possible and must understand that even a single theoretical choice in science might look quite different depending on how one views it philosophically.

That the philosophical issues are of such great complexity, forcing us to look at numerous subtly distinguished alternatives and to engage in an intellectual juggling act of treating several interrelated but distinct issues simultaneously, sometimes makes for hard going in following the argument in some of these essays. The work of the reader is not made easier by the fact that the physical theories used as test cases here are themselves of a high level of abstractness. A full understanding of them would require a mastery of large areas of modern mathematics, and even the nontechnical treatment offered here requires the reader to try to grasp some of the subtle ideas of current mathematical physics. The reader shouldn't be discouraged if working through these essays sometimes seems like heavy going indeed.

Since topics are taken up again and again and treated in different ways and from different perspectives, a good way to approach this book would be to read through the essays from beginning to end, not worrying about the places where full comprehension is elusive. In this way one will gain an overall perspective on the structure of the problems being attacked and on the author's "dialectical" way of approaching them. A rereading of some of the more difficult essays subsequent to gaining this overview will probably find the reader less adrift in the complexities and better able to keep sight of the overall objectives while trying to sort out the necessary but intrusive details which sometimes tend to force one to lose sight of the overall direction in which the essay is going. I realize that this is asking a lot of the reader, and I can only hope that he will find his efforts rewarded by the understanding gained. The author begs forgiveness of his readers for those difficulties engendered by his inability of laying things out in the clearest and most perspicuous way possible. Suffice it to say that strenuous efforts were made to make the argument as straightforward and transparent as possible. I repeat, the issues are complex and difficult, and the gain in clarity which might result from oversimplification would, I am afraid, require paying too high a price in missing out on what are often crucial and fundamental issues.

1. Methodological Conservatism

Don't let yourself get stuck between alternatives, or you're lost.
You're not that strong. If the alternatives are side by side choose the
one on the left; if they're consecutive in time, choose the earlier. If
neither of these applies, choose the alternative whose name begins
with the earlier letter of the alphabet. These are the principles of
Sinistrality, Antecedence, and Alphabetical Priority—there are others,
and they're arbitrary, but useful. Good-bye.
                                    —John Barth, *The End of the Road*

*A core problem in the epistemology of theories of space and time is the
problem of rationally choosing among alternative geometric accounts of
the world all of which agree with one another on the total possible body of
empirical predictions.*

*It has frequently been suggested that our grounds for theory choice
outrun mere conformity with empirical data. Considerations of simplicity
and a priori plausibility are frequently invoked to introduce elements
which might epistemically distinguish between theories which are indistinguishable by means of their conformity or disconformity to observational facts.*

*Sometimes it is claimed that our theory choice is determined by an
additional factor, a factor which is independent of both conformity to
empirical data or to the intrinsic merit of the hypotheses themselves.
Science, we are told, is a* conservative *discipline. Changes in theory are
made only when forced by disconformity of theory with data. And even
then change is always chosen to be minimal change from the preexisting
theory. While we are to eschew dogmatism in science, we are, and ought to
be, methodologically conservative.*

I wish to acknowledge the very helpful comments on earlier drafts of this paper by
Jaegwon Kim, Holly Goldman, and the editors and referees of the *Philosophical
Review*.

"Methodological Conservatism" explores some options concerning just what it might mean to be methodologically conservative. It examines the allegation that doctrines espousing conservatism are simply incoherent. It then asks whether the injunction to be methodologically conservative, even if coherent, could possibly be rationally justified. Finally, it suggests that there is a distinct kind of methodological conservatism, a kind which differs in its epistemic role from the kinds discussed earlier in the chapter, a kind of conservatism rather more profound than those discussed earlier, but also far harder to precisely define and delimit.

The specific question as to whether principles of methodological conservatism could help us resolve the problem of theory choice, where we are concerned to pick, on the basis of idealized observational data, a geometry for the spacetime of the world, is taken up in more detail in chapter 3, section IV, subsection "Variations on Option D."

I

THE aim of this essay is the study of a fragment of a fragment of a program for a theory of the rational acceptance and rejection of hypotheses. After some preliminary skirmishing I will settle down to the brief exposition of one possible version of a principle which I will call the principle of methodological conservatism. I will then explore some arguments for and against the reasonableness of placing such a principle into one's canon of principles for belief. I think that the reader will see in the background of this discussion some very broad issues concerning justification and rationality in epistemic contexts, and I hope that eventually the issues discussed here in reference to the problems suggested by this particular principle will illuminate fundamental issues of rationality of belief in general.

Since the dialectic structure of what follows is somewhat complex, it might be useful here to lay out the structure of the argument. First I consider a number of principles for belief which might be called conservative. I dismiss most as implausible components of anyone's canon for rational belief and then propose a weak principle which is, at least, not as obviously unreasonable as the others. Initially I assume that the principle is one which is "super-added" to a body of canons for belief which function independently of it. I argue that even the weak principle I am considering could play a significant role in our decision-making apparatus when applied as a means of "last resort" when the other principles fail to motivate a decision for us.

Next, I consider the claim that any such principle of conservatism is incoherent or absurd as a principle of decision-making and argue that such a view is not well founded. I then consider the question as to whether or not, coherent as it might be, the principle is justified as a principle of belief. I argue that a utilitarian rationale for accepting the principle can be given, but that, despite this, a persuasive case can still be made for arguing that a canon for rational belief really suffers no great loss when this principle of conservatism is deleted.

Finally, I argue that from a different point of view conservatism may be alleged to play a rationally justifiable role which cannot be dispensed with in any reasonable canon for decision-making. From this point of view conservatism lies at the very basis of any possible structure for justifying beliefs at all. It is argued, from this point of view, that without the assumption of conservatism of this latter kind no justification for belief of any kind could ever get under way. This new view of the role conservatism plays in formulating our belief structure suggests a new rationale for it, a rationale more of a "transcendental" kind than the "utilitarian" justification offered at the earlier stage.

II

What is a principle of methodological conservatism? Basically, the idea is that the very fact that a proposition is believed can serve as warrant for some attitude to be rationally maintained in regard to believing it. The vagueness here is deliberate. More generally, the principle might deal not with propositions believed, but with some aspect of them — the predicates used in expressing them, for example. And it might deal not with the rationality of continuing to believe a proposition, but rather with maintaining some epistemic attitude with regard to these aspects — considering oneself justified in "projecting" predicates which have become "entrenched" by their appearance in previously believed propositions, for example.

A very strong, but limited, principle of conservatism might be the one adopted by those who believe in self-warranting propositions. According to this view there are some propositions, usually immediate reports of sensory experience or "obvious" necessary truths, which we are warranted in believing by the very fact that we do believe them. So believing p is conclusive grounds for maintaining that one *should* believe p. The belief is warranted by the very fact that one has it. Now I do not think that conservatism in its pure form is really in play here, despite the conformity of this principle with our criterion for a principle's being methodologically conservative. For the real warrant for belief in the propositions is their special ("immediately known" or "self-evidently true") nature as private sensation reports, truths of logic, or whatever. The principle invoked serves merely to get us from the realm of is ("I do believe p") to that of ought ("I should believe p"). It is the *nature* of the proposition which is the real ground of its warrant, not the fact that we did, in fact, believe it. Let us eliminate from consideration such principles by insisting that our principle apply to all believed propositions (or their appropriate components) generally, irrespective of any special epistemic status for the proposition due to its "nature."[1]

A clear case of a principle of methodological conservatism, and a bold one indeed, would be the following principle:

> For any proposition, p, the fact that p is believed by x is some warrant for x to continue to believe p.

At least one critic of methodological conservatism, Goldstick, whose arguments we will examine later, took this to be *the* principle of methodological conservatism.[2] He rightly observed that the principle just is not plausible for, according to it, anyone, no matter how irrational or even insane his corpus of beliefs might be, would have on the basis of this principle at least some grounds for holding to his beliefs. But surely not only is believing p not sufficient

1. I think that one can find in the literature proposals for epistemic principles which are limited in their applicability to particular kinds of propositions, but whose nature, and alleged justification, closely resembles that of the general principle of conservatism I will be discussing. See, e.g., the proposed justification of memory beliefs in R. Brandt, "The Epistemological Status of Memory Beliefs," *Philosophical Review* 64 (1955): 78–95; reprinted in R. Chisholm and R. Swartz, *Empirical Knowledge* (Englewood Cliffs, N.J.: Prentice-Hall, 1973).
2. D. Goldstick, "Methodological Conservatism," *American Philosophical Quarterly* 8 (1971): 186.

grounds for believing p, at least in all but the special cases of self-warranting beliefs; in general, believing p is *no grounds at all* for believing p.

But there are more modest principles of conservatism which reasonable men have maintained. The more plausible versions of the principle usually deal with our warrant for not rejecting a belief once it is held, rather than with what we might call "positive" warrant for belief. Let me illustrate with a principle which, I think, is still far too bold to be plausible, although it is more modest in its pretensions than that above:

> Suppose one believes p which serves to explain a wide range of facts. Then one should not reject p, even in the face of disconfirmatory evidence, until one has an alternative explanation of the facts for which p accounted.

As it stands, the principle again seems too immodest, for it views having no explanation at all for some facts as so unsatisfactory that it would have us hold onto an explanation we do have in the face of as much disconfirmatory evidence as we can imagine. And it would have us continue in this path of stubbornness until we have been clever enough to think up an alternative explanation for the facts. But surely when the weight of disconfirmation becomes heavy enough we should surrender. We should admit our inability to explain the facts in question, and we should drop our present theory even though we have nothing to put in its place. Perhaps this principle could be modified to make it more plausible, but I will forgo the attempt here.

Can we find any principle which is methodologically conservative and yet which it is not obviously irrational to espouse? I think that we could find such a principle by exploring in depth Goodman's notion of predicate entrenchment and its role in induction, for I think that the doctrine of entrenchment is one of methodological conservatism. But instead of doing this I will focus on a principle which is easier to state and to comprehend and which is devoid of many of the complexities which make a quick treatment of Goodman's notion impossible. My principle will be genuinely methodologically conservative, not as obviously irrational as those above, and, I think, one which may in fact play a role in scientific decision-making in actual practice. Of course, the rationality or irrationality of the principle, and of those who apply it, if indeed there are such, will be the major issue we will consider.

III

The principle whose rationality I wish to explore is this:

> If you believe some proposition, on the basis of whatever positive warrant may accrue to it from the evidence, a priori plausibility, and so forth, it is unreasonable to cease to believe the proposition to be true merely because of the existence of, or knowledge of the existence of, alternative incompatible hypotheses whose positive warrant is no greater than that of the proposition already believed.

This principle does not commit us to the believability, even in the slightest degree, of a proposition merely because it is believed. Nor does it prohibit us from rejecting a proposition once believed on the basis of disconfirming evidence, even if no better alternative is available. All it commits us to is the decision not to reject a hypothesis once believed simply because we become aware of alternative, incompatible hypotheses which are just as good as, but no better than, that believed.

But how could such a weak principle play any significant role in helping us form a body of rationally held beliefs? One obvious application of the principle is to the problem of the underdetermination of theories by data. This is the situation which arises when, it is asserted, no rational grounds for choosing between two alternative incompatible theories can be found in inductive inference from the data by all reasonable canons of confirmation, rules for inferring to the best explanation, principles of a priori plausibility, and so forth. It is frequently claimed, in works ranging from Descartes through Poincaré to Quine, that if we look at some of the most widely accepted theoretical beliefs in our corpus of beliefs, we discover that there are incompatible beliefs imaginable which would be just as well supported on all reasonable inductive and a priori grounds. And not just as well supported as our current beliefs on the basis of our present empirical data but, demonstrably, just as well supported on *all possible* empirical evidence which would confirm our current beliefs. Sometimes it is alleged that we can actually construct the equally well supported alternative (Descartes's demon, Poincaré's alternative geometries) and sometimes their existence is merely alleged without any attempt being made to say what they would look like.

Now there are a number of classical responses to alleged proofs of underdetermination. One is to maintain that the whole problem is a red herring since it rests upon an untenable distinction between theoretical propositions and "pure observation sentences." If this distinction is a philosopher's illusion, then the alleged demonstration that the alternative theories "save all possible observational phenomena equally well" breaks down. Sometimes it is said that the problem is illusory because one always has rational grounds, based, say, on a priori plausibilities of hypotheses, for choosing one alternative over the others. Sometimes the problem is countered by arguing that truth itself is a relative notion, relative to a "conceptual frame," and that each alternative is "true" relative to some frame and no two to any one frame, and that, furthermore, given a frame we can rationally justify a choice as to which of the alternatives we should take to be "true." Another group of philosophers argues that a proper account of *meaning* shows us that the problem of underdetermination is an illusion, for if each of the alternatives has the same observational consequences, then the "alternative" hypotheses really are not alternatives at all — they are all merely the same theory multifariously rewritten in misleading ways. From this point of view the illusion of underdetermination is a consequence of the fallacy of equivocation. Finally there is the "permissive" view of rationality, which argues that since belief in each of the alternatives is equally well *permitted* (although not obliged) by the evidence, belief in any one is rational although belief in more than one plainly is not.[3]

At this point, I will beg the reader's indulgence and ask him to go along with me in assuming (1) that the problem of underdetermination is real in that the alternative hypotheses really exist and really are incompatible with one another — that is, they are not different expressions of one and the same theory; (2) that it does make sense for us to talk of "all those principles of rational belief except conservatism" and of "incompatible alternatives all supported to an

3. For an extended discussion of some of these issues, both in the general methodological context and as applied to the case of our geometric theory of spacetime, see the author's *Space, Time, and Spacetime* (Berkeley: University of California Press, 1974), esp. 88–146.

equal degree by all possible evidential data utilizing these rules";
and (3) that we will not be satisfied with that version of a "permissive" use of the notion of rationality which tells us that when we
discover a new hypothesis just as warranted on all positive grounds
as a hypothesis we currently believe, it would be equally rational for
us either to go on believing our present hypothesis or to transfer our
belief to the newly discovered alternative. This is a lot to swallow,
but I think that we will learn much by putting these issues temporarily to the side and pursuing the issue from here. We will return to
some of these questions shortly.

Even if the reader is skeptical about the very existence of the kind
of underdetermination we have been describing, he is likely to
agree that a second kind of underdetermination does exist. Let us
call this "transient" as opposed to "radical" underdetermination.
Here we allege merely that there can be incompatible alternatives
between which no rational choice can be made on the basis of a
priori plausibilities, strength, simplicity, inductive confirmation,
and so forth, *relative to present empirical evidence.* In this case
future data might very well make one of the alternatives uniquely
most preferred on the basis of these other "non-conservative"
grounds. It is only now, given our present evidential basis, that the
theories are underdetermined relative to current observational
considerations. Even those skeptical of the very possibility of radical underdetermination are likely to admit that transient underdetermination is a fact of epistemic life.

How can a weak principle of conservatism of the kind described
above aid us when we are faced with a case of underdetermination?
On reflection I think that we can see that there are two distinct
problems of underdetermination:

(1) Suppose we already believe a hypothesis. Then we discover new
hypotheses which are just as well warranted on the basis of all possible
(or current) evidence as the one we now believe.

(2) Suppose we are trying to formulate a novel hypothesis to account
for newly discovered empirical facts. How can we choose among the
many "observationally equivalent" hypotheses all of which are equally
rationally believable on the basis of the data?

The principle of conservatism clearly resolves our difficulties in
the first case, for on the basis of it we simply go on believing what we
always have believed. And we do not diminish our confidence in

the correctness of our belief a jot just because we realize that alternative beliefs, no better than what we presently believe on the basis of all principles other than conservatism and worse in that respect, exist.[4]

The principle can be argued to play a subsidiary role in the second kind of case as well, although here the argument is more tendentious. The argument goes like this:

We have to choose among alternative hypotheses all of which seem equally warranted on the basis of reasonable inference from the evidential data. But suppose one of the hypotheses is "more in conformity with preexisting theory"—more "like it in structure," say, or more "conceptually continuous" with it. Then that hypothesis is to be preferred. Notice that we are assuming that this preferential support given one of the hypotheses by the "background theory" cannot be traced back to any greater a priori plausibility for the hypothesis in question, nor to any superior inductive support for it from the evidence basis for the background theory—for we are assuming that the alternative hypotheses are on a par in these respects. Rather, the claim is that insofar as the preexisting theory is the result of choices that outrun those rationalized by a priori and inductive considerations, the selection we have already made— say on the basis of conservatism—can be "projected" into a method for selecting among novel hypotheses.

There are two major points to be made here.

(1) I doubt that a plausible case can be made out that we do, in fact, make our decisions in the case of a choice among novel "undetermined" alternatives by comparing their conformity to preexisting theory. There is nothing, as far as I can see, in the physical theory which existed prior to 1917 which would lead one to prefer a theory of curved spacetime to one with "universal forces," to give just one standard example of underdetermination.[5] In general, I

4. Throughout this essay I assume a simple classificatory model of belief in which propositions are either believed or disbelieved. It would be interesting to pursue the form principles of conservatism might take in a schema which allowed for degrees of belief and in which, perhaps, being believed counted as *weight* toward being believed. But I will not explore this issue further here.

5. A defense of this, admittedly rather bold, claim is clearly outside the scope of this essay. For some relevant arguments, see Sklar, *Space, Time, and Spacetime,* 88–103 and 113–146. See also "Facts, Conventions, and Assumptions in The Theory of Spacetime," chap. 3 in this volume.

am dubious that reference back to the structure of preexisting background theory will usually uniquely select as preferable one of the numerous alternatives in a case of underdetermination with respect to novel hypotheses.

(2) *If* this method ever is applicable, however, then our principle of conservatism does have a role to play. Suppose we choose H_1 over H_2 on the basis of greater conformity to background theory. Someone might then reply: "But since there are alternative background theories all as warranted as that we do in fact accept, the choice of H_1 based on the background theory is still arbitrary and unwarranted." It is now that we invoke conservatism. Since we do believe the background theory we do have, and since H_1 is preferable relative to this background, and since our principle of conservatism tells us to prefer the background theory we do have to any of the alternatives we might have had, we *are* warranted in accepting H_1, even if H_2 would have been preferred had the background theory been one of the "underdetermined alternatives" to the theory we did already accept.

Obviously the application of the conservative principle is simpler and more decisive in the case where we are concerned with sticking with a hypothesis which we already do believe than it is in the case of selecting from among a set of novel hypotheses. So let us focus on this situation. Is the adoption of the rule justified or reasonable even in these cases? Clearly the rule does resolve a dilemma for us — it tells us to stick with the theory we have and not to drop it for one of the newly discovered alternatives nor to lapse into skeptical suspension of belief. But is conservatism itself warranted?

IV

Our first task will be to dispose of a poor argument against conservatism. This is the claim that the principle is in some sense logically incoherent or absurd. We will shortly examine more subtle arguments for being skeptical about the principle.

A typical argument to the effect that conservatism is absurd is Goldstick's.[6] Goldstick deals with a principle of conservatism far stronger than mine, but his major argument goes through against my modest principle if it works at all. Further, Goldstick believes that we must either accept conservatism, be caught in a vicious

6. Goldstick, "Methodological Conservatism," 186–191.

circle of justification, or be committed to some version of foundationalism in epistemology—if we are to avoid skepticism. This may be right but I will not pursue that issue here.

Goldstick's argument against conservatism is this. Conservatism could lead us to the position of having two individuals, possessed of all the same evidential facts, both of whom we adjudicate as being rational in their beliefs, despite the fact that they hold incompatible beliefs. And that, says Goldstick, is absurd.

But is it? Consider two societies, A and B, organized on different social systems. Suppose their systems serve equally well in accomplishing identical social goals. The people in society A argue as follows: "Our system has a certain efficacy in obtaining our social goals. The existence of society B shows us that an alternative system of social organization would do the job equally well. But since we are organized according to our system we are justified in keeping it, for stability of a social system is itself a desideratum, and one should change systems only if there is a *better* system to change to. There is no point whatever in changing to a new system which is merely as good as the one you've got." Clearly the members of B argue analogously. Surely there is nothing absurd about a political conservatism which is that modest.

But isn't the case different where it is the rationality of *beliefs* which is at issue? For haven't we neglected the issue of *truth,* and isn't that what distinguishes the epistemic from the political case? By hypothesis, since the alternative hypotheses are supposed to be incompatible, both parties cannot be correct in their beliefs, but there is no sense in which either society A or B can be said to be "incorrect."

But the invocation of truth will not serve to show epistemic conservatism incoherent or absurd. Surely we do not want to declare a man irrational in his beliefs simply because they are false. Nor do we want to be committed to the position that whenever two persons have incompatible beliefs one of them must be irrational. Consider the case where they have different observational evidence to go on. Surely both can be rational although at least one must be wrong. Notice that this is *not* to espouse the kind of "permissive" use of the notion of rationality mentioned above. It is one thing to assert that two different individuals with their different starting points may be equally rational despite the fact that they accept incompatible hypotheses. It is quite another matter to claim that a given individual may be rational no matter which of a number of incompatible hypotheses be adopts.

But, the reply will be, Goldstick is considering the case where both parties have the same evidence. Now it would be wrong here to reply by taking what the parties do in fact believe to be a difference in their "evidence." And we need not make this move. The fact is that the parties do differ in their "total state," they differ in which hypothesis they believe. And if it is *logically coherent* to maintain that two parties who differ in their evidence can both be rational even though their beliefs are incompatible, then it is logically coherent to maintain that they are equally rational in their incompatible choices even though their different warrant for their beliefs is based solely on the fact that they *do* believe different incompatible hypotheses and each adopts a principle of conservatism which tells him not to change his belief unless there is some reason to do so.

To charge conservatism with incoherence might be to rely on a claim that any rule for rational belief is incoherent which declares that there can be equally rational people with incompatible beliefs. But such a principle of coherence would make a mockery of our usual, quite reasonable belief that a man can be rational but wrong. And to adopt the line that the principle is logically incoherent because it allows us to declare believers in incompatible propositions equally rational even if their beliefs are based upon the same evidence and have the same a priori plausibilities would simply be to beg the question by assuming that these are the only rational bases for decision-making and that conservatism is irrational.

V

But a principle can be logically coherent and yet unjustifiable or even grossly irrational. Is conservatism a justifiable principle for decision-making? It is to this question that we now turn.

In trying to decide if conservatism is a justifiable rule for rational belief we are in deep water, for what is required is a general account of justification in epistemology, and that is clearly far too gigantic a topic for the compass of this paper. But I will make some comments on justification in general, sketchy as they must be. Let us consider a number of "modes of justification for rules for belief" and see how conservatism fares according to each approach.

(1) *Justification by Intuition.* A rule is justified if it is intuitively obvious that it leads to truth or, perhaps, at least to probable truth.

This is certainly the view that some have held about the justification of the rules of inference in deductive logic. Others would extend it to rules of "intuitively sound" inductive logic. It hardly seems to help in our case. Surely there is no intuitive guarantee that the theory we do believe is more likely to be true than any of the equally warranted alternatives which we reject simply because we do not, in fact, believe them.

(2) *Justification by Codification of Practice.* From this point of view what we do to justify a rule is this: we look at cases of inference we accept as justified. We then seek general rules which subsume these particular cases of accepted inference. Of course, in the process we may discard some particular cases we took as justified as not being really justified, if, for example, they fail to fit a simple set of rules which covers all the "best" particular cases, or if on reflection generated by our search for general principles we change our minds about the acceptability of some particular inference. Thus Goodman has characterized as the problem of induction this: to find the rules which subsume intuitively accepted cases of inductive reasoning. The rules are justified by the fact that they codify so many particular intuitively accepted cases, and new particular inferences are justified by their conformity to the rules.

I do not think we need pursue the obvious line of dialectic which arises in a general critique of this point of view, for it hardly seems to do much good in the case at issue. Here we do not have the interesting problem (which appears in deductive or inductive logic) of finding descriptively adequate rules to cover the particular cases. And surely in this case there is just as much skepticism about the application of the rule in particular cases as there is about the justifiability of the rule in general. Someone who is skeptical of the rationality of some particular decision based on conservatism is hardly likely to be appeased by reference to the conservative principle, and it is also doubtful that anyone skeptical of the rule's rationality would be reconciled by being shown that we sometimes do behave in a methodologically conservative way.

And if we consider the line of some exponents of "naturalized epistemology" that all we can do as philosophers is codify "natural" practice, we need only note that not only is acting conservatively in epistemic cases natural; so is being skeptical of the rationality of one's action when it is pointed out to one that that is what he has been doing!

(3) *Justification by Appeal to Higher Rules.* Sometimes we justify accepting a rule by showing that it is derivable from some higher or more general rule of the same "type." Derived rules of deductive inference, for example, are justified by their derivation from the more fundamental rules.

But our rule of conservatism is itself of such great generality and simplicity that it is dubious that it can be given any interesting kind of justification by subsumption under some more general principle of the same general sort.

(4) *Justification by Empirical Grounding.* A version of "naturalized epistemology" distinct from that discussed in (2) above tells us that the justification of a methodological rule involves demonstrating that it is a "good guide to truth," where in this demonstration we are allowed to use any and all of our best current theory about the nature of the world.

Now in some of its versions this "scientific" theory of rationality comes dangerously close to resembling what Hume criticized so effectively as the inductive justification of induction. What good is the rationalization of a rule which rests upon scientific beliefs which presuppose the use of the rule for their warrantability? Later we shall see that another variation of this approach may be immune to the charge of vicious circularity and we shall see that some version of methodological conservatism is an essential component of this account of rationality.

There are some applications of principles of conservatism toward which this approach to rationalization might have at least initial plausibility. Suppose we have a principle which tells us, given that we believe H_1, to continue to believe H_1, even in the face of the discovery of some new hypothesis H_2, equally warranted on all grounds other than conservatism at least with respect to *current* empirical evidence. Here we are considering the application of conservatism to a problem of transient underdetermination. We *might* have good scientific reason for believing this: Whenever such a case has arisen in the past, we have discovered, when the new empirical evidence which differentially supports H_1 and H_2 has come in, that more often than not the original hypothesis thought up by the scientific investigators has turned out to be the better supported one once the new evidence became available. Then we might have "inductive" and "scientific" reason for rationalizing the use of the conservative principle at least in cases of transient underdetermination. That we actually *would* have such grounds, or that we do, seems dubious—to say the least.

Of course, in the case of radical underdetermination, no such "empirical" warrant for the application of the conservative principle could be possible, for here, by hypothesis, no further accumulation of empirical data could ever show us that our application of the conservative principle did indeed lead us to make the "right" choice.

As I noted above, we shall see that while appeal to best available scientific belief cannot rationalize a principle of conservatism, something like a principle of conservatism will play a crucial role in justifying the principles which rationalize inference by appeal to current belief. But I will postpone discussion of this point for the moment.

(5) *Justification by Appeal to Means and Ends.* Science is a purposive human activity. The rationality of actions is to be adjudicated in terms of their suitability as means toward the end of the action. This is just as true when the end is "coming to believe a body of rationally accepted scienic hypotheses" as it is in the case of actions whose purpose is a more clearly specific and delimited "practical" end. So we should at least consider the possibility of justifying the rules for belief in terms of their suitability as means for obtaining the kinds of beliefs which are the ends of scientific research.

Now it is obvious that a full discussion of this point of view is beyond the scope of this essay. So I will make only a few remarks suggestive of the way such a justification would go when the rules in question are not the conservative principles we are dealing with. Then I will discuss how such a means-ends justification applies to the conservative principle and just how this justification in this particular case is rather unlike its application in some other cases.

Why should we use inductive logic? Well, we are aiming at truth in scientific research, and confirmation, if there is such a thing, is a "partial warrant" for believing a hypothesis true. The warrant for confirmation is its role as means toward the end of *safety* in our beliefs.

Why should we adopt the rule "Accept the strongest hypothesis available, all other factors being equal"? Well, because, as Popper has so vigorously emphasized, we *want* to believe in strong hypotheses, for it is they which have the maximum predictive value given that we do believe them. Similarly, we might justify the rule of picking the simplest available hypothesis in terms of the desirability of simple scientific theories.

How would conservatism fare under a program of rationalization? I think its rationale is interestingly different. Here the argument might go like this: Stability of belief is itself a desirable state of affairs and an end to be sought. A hidebound refusal ever to change one's beliefs is nothing but irrational dogmatism. But the desire to maintain the beliefs one already has *unless there is some good reason to change them* is as rational as the programmatic commitment to maintain one's social institutions unless there is some reason to revise them. In both cases the rationale is the same — change itself is an "expense." It requires effort, energy, will, communication, and so forth. And effort uselessly expended is effort irrationally expended. So unless there is good reason for change, things should be left as they are; and this holds as much for scientific beliefs as for social institutions.

We should note a crucial difference between the rationales offered for confirmation, strength, and simplicity and that offered for conservatism. All the former cases rely upon a justification which utilizes some feature of the hypothesis in question — its "degree of probability" relative to the evidence, or its strength, or its simplicity. But the rationale for conservatism is totally independent of any facts about the nature of the hypothesis to be conserved or about its relation to the evidence. The "utility" involved in the rationalization of conservatism is the utility of the conservative principle as a rule for belief in general, and not the utility (expected or intrinsic) of the belief in the hypothesis which is adopted when the rule is followed. We should be conservative because conservatism itself has utilitarian value, and not because we believe that we will increase the utility of our belief set, or even that we will probably do so, if we follow the conservative rule. There is a utility of *methods* which is not assimilable into utility or expected utility of the results of applying the method.

So we should be conservative, in my weak sense, because it is a waste of time, effort, and energy to change your mind for no good reason. And since, beliefs being our only guides to prediction and action, it is generally preferable to have some belief than to have no belief at all, it is even irrational to drop a belief held, thereby replacing belief by skepticism, unless forced to do so by contravening evidence or something of that sort. The mere existence of alternative beliefs equally warranted on all other grounds but conservatism is not enough to warant dropping a held belief. It is as simple as that. Note also that this justification for conservatism holds in the cases of both radical and transient underdetermination.

VI

The reader initially dubious about the rationality of a principle of conservatism probably remains skeptical at this point, so it will be helpful to explore the grounds for his skepticism a little further.

We have seen that two grounds for skepticism might be that (1) the principle of conservatism might lead one to prefer a false hypothesis to a true alternative, and (2) if we accept conservatism we may be led to attributing rationality to two persons who espouse incompatible hypotheses. But if those were one's grounds for rejecting conservatism, one should, by parity of reasoning, reject any principle which allows one to infer from the data to hypotheses not logically entailed by it. I will assume that the reader is not that hard-nosed a skeptic.

I think that many are skeptical of conservatism because of its "arbitrariness." "Look," they say, "it is merely accidental that you came upon one of a set of equally warranted hypotheses first. Yet conservatism tells you to prefer the firstborn to the others. But any of the alternatives could have been the firstborn, so your rule of primogeniture inserts an unacceptable degree of arbitrariness into our canons for rational belief."

Maybe something can be made of this, but it is hardly persuasive as it stands. What evidence we accumulate first, it may be argued, is just as "arbitrary" as which hypothesis to explain it we come upon first. Would we reject Newton's inference to his theory of spacetime as irrational just because it was the facts about absolutely accelerated objects which came to his attention and not the results of the Michelson-Morley experiment which were not yet available, but if the latter had come first he would never have inferred to the spacetime theory he did accept? And if it is replied that what evidence comes first is a natural fact of the world, and not a matter of "mere choice" on our part, isn't the clear reply available that it is just as much a part of "nature" which hypotheses come first to mind as it is which facts come first to observation?

I think that a more subtle and enlightening case for a kind of skepticism with regard to conservatism can be found by exploring our attitudes toward the beliefs of ourselves and others when these beliefs are arrived at by using a variety of principles for belief. If we utilize a number of different rules for deciding upon our corpus of accepted propositions we may very well find ourselves in the position of (1) believing a hypothesis, (2) rationalizing our belief by means of a principle for rational belief which we accept, and yet (3) at the same time wishing to indicate to ourselves and others the fact that our accepted hypothesis is preferable to some alternatives only on the basis of a limited subset of our principles of inference — and perhaps only on the basis of a principle to which, in some sense, we give low priority.

Thus, I think a position like the following is both internally coherent and not unlike the attitude we frequently do take with regard to theoretical beliefs: "I believe in curved spacetime, and in the theory that free particles travel timelike geodesics and light rays null geodesics in this spacetime. Now I fully realize that an alternative total theory (universal forces) would explain all the possible empirical data that my theory explains and would, in fact, explain it equally well. Further, I believe this alternative theory to be as strong, simple, a priori plausible, and so forth, as my theory. Yet I believe my theory and disbelieve the alternative. I did believe it, prior to realizing that there was an alternative, and I see no reason to give it up. There is no evidence against it, and the alternative theory will not do any better job on the positive evidence that there is. So, on the basis of conservative principles, I believe it rational to stick to my original hypothesis.

"Yet I hardly want to say that I *know* that spacetime is curved. And not simply because I am not *certain* that it is, for I do not take certainty as an essential prerequisite for knowledge. I deny knowledge to myself because I realize that my preference for this theory over its alternatives is based solely on conservatism. The theory I do accept is no better confirmed by the data, is no stronger or simpler or more a priori plausible than the alternatives. I mark out the 'weakness' of the grounds for the preference I do have by attributing to myself rational belief but denying to myself knowledge.

"Notice also that the principle I have invoked, conservatism, is of low priority — in the sense that if any of the alternatives excelled my present belief in any way (degree of confirmation, simplicity, and so forth) I would accept the alternative rather than what I do accept. My principle of conservatism is a principle of 'last resort' in my decision-making canon. That is another reason why I refuse to countenance as knowledge those beliefs the rationale for which invokes the conservative principle."[7]

Even if we believe that the principle on which we made our decision is the *only* working principle for rational decision-making in science, we may still wish to qualify our attitude toward beliefs accepted on its basis as "rational nonknowledge." I think that something similar to this is what is going on when Popper tells us that it is rational for us to "accept" the strongest corroborated theory available on "pragmatic" grounds, but at the same time tells us that we have no reason whatever for *believing* such theories to be true. Popper believes that there is no such thing as inductive confirmation. He believes that logical conformity to the available evidence and strength of hypothesis are all we have to go on in our scientific decision-making. He believes that it is rational to accept hypotheses on these grounds. Yet he is so anxious to remind us that these are *not* the grounds for belief which confirmation would supply—if there were such a thing, which, he believes, there is not—that he is driven to the device of disassociating scientific acceptability from rational believability for fear that the latter term would suggest that his rules of acceptance were something other than what they actually are.[8]

We know a number of reasons why one might want to say that another holds a true belief on rational grounds but fails to have knowledge. These involve the well-known cases of his use of false lemmata, our knowledge of further evidence not available to him, and so forth. What I am suggesting here is that we sometimes take this attitude toward others and toward ourselves for another reason. We may have a rather large collection of what we take to be canons for belief. But we may wish to note the fact that some of our beliefs require the application of rules from this canon which other beliefs do not. We may wish to indicate the fact that we have had to

7. It is an interesting question whether conservative considerations could ever outweigh those of confirmation, simplicity, strength, etc. It is certainly the case that the history of science seems to show us that practicing scientists do let conservatism outweigh these other considerations. Whether they should is another matter.

8. K. Popper, "Conjectural Knowledge: My Solution of the Problem of Induction," *Revue internationale de Philosophie,* 25e année, no. 95–96 (1971), fasc. 1–2. Reprinted in his *Objective Knowledge* (Oxford: Oxford University Press, 1972). See especially p. 22 of the reprint.

invoke these special rules in a particular case, and that for that reason the beliefs we hold in this case are not as "well founded" as those beliefs we can support without reference to these special rules. The man who accepts methodological conservatism may want to say that the beliefs he holds only on the basis of the invocation, among other rules, of his conservative principles are rational beliefs but not "knowledge." Alternatively, he may indicate the weakness of his grounds for belief in this case by calling his attitude toward the hypothesis one of "acceptance," rather than one of belief.

VII

But let us pursue the skeptic's doubts a little further. We have noted two cases where conservatism may be invoked: those of radical underdetermination and those of transient underdetermination. About the former cases the skeptic is likely to argue that one of the positions we dismissed earlier (there are no true radically underdetermined hypotheses — either because they are merely expressions of the same theory or because the whole puzzle rests on an untenable observational theoretical dichotomy) begins to sound ever more persuasive the more one sees what is required for decision-making if radical underdetermination ever did arise. But even those dubious about the possibility of radical underdetermination are likely to admit that the transient case is a reality. How would someone dubious about the rationality of conservatism tell us to proceed in these cases?

I think an argument might go like this. Suppose we believe H_1 and then discover H_2 which is just as plausible, on all but conservative grounds, as H_1 relative to present evidence. What should we do? The conservative tells us that considerations of utility recommend our sticking to our present belief. But that is *not* necessarily what utility does necessitate. What we should do depends, first of all, on the relevant utilities in the particular case of not believing anything, believing something and having it be true, and believing something and having it be false. Just how important is it (on either "practical" or "purely scientific" grounds) for us to have some belief or other? If it is not all that important, then the thing to do is to admit that one just has no idea which hypothesis is true and remain in a skeptical withholding of judgment until further evidence is in.

If it is important for us to believe at least one of the competing hypotheses, even at the risk of selecting the wrong one, then we should choose one for tentative belief. But in making this choice there is nothing in the way of "rationality" which makes the first hypothesis to come along preferable to one of the others. The kinds of methodological utility conservatism invokes in its justification carries no weight in deciding upon the rationality of belief. Of course, it might be less *convenient* to adopt the principle "Choose the second hypothesis to come along" in place of the conservative principle, but such matters of convenience are irrelevant to questions of epistemic (as opposed to pragmatic) rationality. To mark out the "weakness" of the grounds for our decision we might be wise to speak of "adopting but not believing" the theory we do, in fact, choose.

Alternatively—and this move is applicable even in the radical underdetermination case—one could opt for the "permissive" notion of rationality. From this point of view each of the incompatible choices is equally rational, rationality requiring only a choice which is permitted, rather than one which is obligatory.

VIII

One direction one could follow up at this point would be a careful examination of the alleged distinction between those principles having "merely pragmatic warrant" and those "epistemically justified." One would like to see in careful detail just how conservatism, skepticism of the kind we have been opposing to conservatism, the distinction between a proposition's being believable and being merely acceptable, and permissivism each fare depending upon one's attitude toward the adequacy of a fully "pragmatic" approach to the warranting of belief.

Instead I will take a different tack here. What I want to argue is that there is a plausible case for the following thesis. Conservatism is not just a minor "last resort" principle invoked only when all other principles have failed to do the selecting job for us. Conservatism is, in fact, so deeply and pervasively embedded in our schema for deciding what it is rational to believe that once we have seen the full role that it plays we are likely to reject the alternatives to it of skepticism, which tells us to withhold belief from any of the alternatives, of permissivism which tells us it is all right to pick any one we choose, or of speaking of our choices as being "adoptions" rather than beliefs.

Thus far we have been dealing with conservatism as though it were essential to decision-making only at a rather superficial level. Decisions based on conservatism, we have been assuming, come into play only after we have pushed our other apparatus for decision-making to its limits and found that we wished to make additional decisions which had no basis for rationality on these "primary" grounds. If this were the case, then to deny the rationality of conservatism would be to deprive ourselves only of a very limited fragment of our decision-making apparatus. But is this so?

Let us reflect on an attitude toward justification we considered earlier. In (2) we asked whether conservatism could be justified if we accepted the view that justification, in general, is always based on a body of "best accepted scientific belief," and we saw that invoking this model of justification helped us little in rationalizing our utilization of conservatism.

But let us now reconsider the view, shared by a number of philosophers of otherwise radically divergent casts of mind, that *all* epistemic justification is relative to an assumed background of believed theory. From this point of view there are no "foundations" to knowledge, in the sense of points back to which we can trace our beliefs in a justificatory way and which themselves require reference to no other beliefs for their justification. These "ultimate beliefs," it is said on the thesis we are pursuing, were sought as the basic points on which a total "global" justification of our belief structure could be founded. But, the view continues, such justifiers-not-themselves-in-need-of-justification are a philosopher's myth, a myth engendered by the mistaken view that there could be, or need be, any such "global" justification for our beliefs taken "all at once."

Instead, it is argued, we must realize that all justification is "local." We justify the beliefs we take to be in need of justification "one at a time," using all the resources of our unchallenged background belief in the process. Such "local" justifications are the only justifications of which we can make sense, for all justification requires a body of unchallenged background belief, and we never could justify our totality of beliefs "all at once." But, the position continues, such "local" justifications are all that is necessary to make sense of rationality in science.

For example, in talking about the rationality of belief we have used the notion of evidence. But what are evidence propositions?

Not, this view asserts, some ultimately self-warranting reports of private awareness, or anything of the sort. They are, instead, unchallenged background beliefs. Now some philosophers who hold this view may be "foundationalists" in a weak sense — that is, in the sense that they would restrict the class of evidential propositions to a proper subclass of our total beliefs (for example, Neurath's protocol sentences or Popper's spatiotemporally restricted singular existence statements). Others deny even this weak version of foundationalism, allowing any currently unchallenged scientific belief, no matter how general or theoretical, to play an evidential role. But all agree that the justificatory process proceeds a step at a time, and all agree that justification is, and needs only be, relative to currently unchallenged scientific belief.

Again, we earlier assumed various methods of justification which we imagined as having been applied before conservatism was brought into play — such as deductive and inductive inference, and so forth. Now the claim will be that there are no such principles which have their justification in any "self-warranting" way. Rather, it is claimed, the rules of inference which we may challenge in some particular case are, once again, justified only by reference to our currently held body of unchallenged belief. Perhaps the extreme version of this is Quine's view that even the principles of inference we call deductive logic are but the application to new cases of the truths we hold at the core of our present empirical theory of the world, truths held so firmly that we mark them out with the honorific title of "logical truths." But, it is claimed, these truths, and those which ground inductive inference if, indeed, there is such a thing, are nothing more than fragments of our momentarily unchallenged belief corpus. We are warranted in using them to decide upon the acceptance and rejection of new hypotheses only because their own status is at the time not under scrutiny or challenge.

One standard objection to such models of justification and rationality amounts, more or less, to reworkings of old objections to coherence theories of truth. Here they are taken to be objections to a "coherence" theory of the rationality of belief. If all justification is "local" and relative to background theory, isn't it possible that our whole system for accepting and rejecting novel hypotheses amounts to just an apparatus for propagating new error from old? And just as the objection to the coherence theory of truth was frequently couched in terms of the alleged possibility of alternative, incompatible but equally coherent total bodies of belief, now the point is raised that without some places at which the body of belief *as a whole* is rationalized by reference to something outside the corpus of belief, can't we imagine wholly incompatible total belief structures all equally "rational" from the point of view of "local" justification?

I think that one can now see how a retreat to conservatism at a deeper level might appeal to an exponent of the thesis of "local" justification. The principle which allows us to ignore the possibility of a wholly alternative global corpus of belief is the principle that the one we have is the one we have. Since justification and its opposite, challenge, are only local and relative to an assumed background, we need not concern ourselves with such hypothesized total alternatives. For the complete and adequate answer to the challenge to the rationality of beliefs as a whole is that without some reason for change, sticking with what you have is the only rational thing to do. And there never can be a good reason for doubting our beliefs in toto, for *real* doubt, like real justification, is always itself grounded in an unchallenged background theory and is always "local."

If the account of justification just offered is correct, and if it is correct that our principle of conservatism plays a crucial if implicit role in such a justificatory structure, then it is clear that a new aspect has been added to the earlier, inconclusive debate I have outlined between those who would install conservatism as a canon of rational belief and those who consider it dispensable. If the only place conservatism played a role was as a principle of "last resort" in a limited number of cases of radical or transient underdetermination, then the position looks plausible which advocates that we drop the conservative principle and remain skeptical, or that we take any of the alternative decisions open to us as permitted and hence rational, or that we accept the recommendation of the conservative principle but consider our decision to be one of accepting but not believing the hypothesis it favors. But if all our beliefs, even those founded most directly on the immediate data and those inferred from it by our most sacred canons of deductive and inductive inference, rest implicitly upon acceptance of conservatism as a principle for belief, then, I think, we would be far more reluctant to look upon conservatism as at best an easily dispensed-with canon of belief. Do we really wish to remain in skeptical indecision about *all* our best-supported scientific beliefs? Or take them as, at best, one among many "permitted" hypotheses? Or take it that we never really have the right to believe any scientific hypotheses, but at most have the right to "accept" them?

We earlier looked at a "utilitarian" argument for conservatism and we saw that an equally good "utilitarian" argument for dispensing with it seemed available. Our argument for conservatism now is more "transcendental" than "utilitarian." If this last account of rational belief and of conservatism's role in it is correct, then without conservatism there could be no rational belief at all.

IX

I do not know if the above is an adequate account of the position of any actual philosopher who propounds a model of rationality of the kind we have been considering. Without our being offered more than metaphorical accounts of stress-relieving spiders saving their webs of belief at minimal expense of effort, or clever sailors rebuilding their fleets without the aid of dry docks or home ports it is hard to comment in detail on full-blown positions of this kind. I believe that there is no adequately worked-out "localist" theory yet available for our detailed critical examination. What is needed is a thorough examination of what the fundamental premises and full internal dynamics of such a model of justification should look like, a careful investigation of the rationales, of whatever sort, that could be offered for each component of the scheme, and a hard look at the obvious critical objections which could be launched at such a model for rational belief — for example, a thorough examination of the alleged "arbitrariness" of beliefs so arrived at and a careful discussion of the ways in which a scheme which invokes conservatism in such an essential way can avoid falling into a dogmatic refusal to be prepared for, and to take seriously, radically different conceptual approaches which may have been ignored in past science and which may be unfairly ignored in the future if we rely on a methodology which puts so much weight on reliance on past theory to project to future hypotheses.

What I wish to suggest here is only this: If all justification is "local" and at best relative to a background of currently unchallenged theory, then something like a principle of conservatism of the kind we have been discussing may play a crucial role at the very center of epistemic justification. The role it plays may not be that of one more principle of justification super-added to others which are more fundamental and are to be applied first, but rather that of a foundation stone upon which all justification is built. And if that is true, sacrificing conservatism on the altar of skepticism, on that of "permissivism," or on that of acceptance as opposed to belief may be giving up a lot more than appeared at first sight.

2. Saving the Noumena

I want to consider another defense which I find most unsatisfactory. It might be called the *positivist defense;* and some philosophers seem to think it is the only defense available to someone who wants to claim that the apparent conflict between two theories is merely verbal. According to the positivist defense, *whenever* we have two theories that have all the same observational consequences, any apparent disagreements between the two theories are merely verbal ones. Call this the *positivist principle.* . . . For instance Sklar's discussion of conventionalism about geometry seems to presuppose this view. In all the standard cases of alternate geometries (plus compensating adjustments elsewhere in the physical theories), the geometric objects of one theory are definable out of the geometric objects of the other. Sklar obscures this fact by comparing the conflict between alternative geometries to the conflict between the normal world-view and Descartes' "evil demon" hypothesis; but this latter example is one where the objects of one theory are clearly not definable in terms of the objects of the other, so only by some form of the positivist principle could one claim that the conflict between those theories was purely verbal.

Hartry Field[1]

Faced with apparently alternative theories, all equally well saving all possible empirical phenomena, the positivist resolves the dilemma by declaring that each of these alternative "theories" is, rather, an expression of one and the same theory differently verbalized. The realist denies that empirical equivalence is enough to constitute full equivalence for theories. Sometimes, however, what seem to be incompatible theories are, surely, merely nominal variants and so equivalent to one another.

My thanks to Steve White for the title and to Tim McCarthy for many useful discussions.
1. H. Field, "Conventionalism and Instrumentalism in Semantics," *Nous* 9 (1975): 392 and footnote 7 on p. 404.

This piece explores one approach that the realist might take in attempting to characterize a nonpositivist notion of the equivalence of theories, and explores some of the difficulties the realist might face in attempting to integrate such a notion of theoretical equivalence into his overall realist understanding of theories and their role in science.

First, it is suggested here that the realist may have to swallow more than he would like of such traditional positivist notions as a hard observational/nonobservational distinction in order to formulate coherently his criterion of theoretical equivalence. Next, it is suggested that this realist notion of theoretical equivalence commits the realist to a specific view of the way in which meaning accrues to the theoretical terms of science.

Once one allows for inequivalent but observationally equivalent theories, as the realist wishes to do, one then must face up to the threat of epistemic skepticism which immediately arises. I explore what seems to me the most common realist response to the skeptical challenge, the view that one ought to take as rationally believable the minimal *realist theory adequate to account for the data. Some problems with this view, both internal ones which make its characterization a matter of some difficulty, and far more profound ones endemic to the epistemological threat to realism in general, are explored.*

Finally, it is argued that the realist is also obliged to answer some fundamental questions regarding his notion of the explanatory role of theories, questions which are made especially difficult for him, I believe, when the full consequences of his most natural choice of a notion of theoretical equivalence is made. That is, I argue that adopting the most common notion of theoretical equivalence espoused by recent realists forces upon them a particular understanding of the way in which meaning accrues to theoretical terms, and that this view of the meaning of theoretical terms is the one which makes it most difficult for the realist to respond to traditional positivist objections to the realist's notion of the value of theories in explanation.

I

M OST, but not all, philosophers would agree that surface differences between two theory presentations might mask underlying identity of theory. Sometimes the verbal presentations of a theory may differ in a way which does not even suggest incompatibility, for example in the case of the presentation of a theory in two different natural languages. Few would want to argue that Newton's *Principia* in Latin does not present the same theory as that work in English. More interesting are those cases where only one natural language is involved, and where, at least on the surface, some sort of incompatibility or other seems present, but where many, if not all, would agree that this surface incompatibility is deceptive.

Are our theories of the world really incompatible if they differ simply in that you talk of temporally extended physical objects whereas I deal with time-slices as my basic particulars? More radically, even if it is a case of a thing theory versus a theory which takes spatiotemporal regions as primitive and treats things as substantival predicates of such regions, many would certainly argue that we have here a case of theoretical equivalence. Hardly anyone would deny that the Schrödinger and Heisenberg "representations" are, indeed, representations of one and the same theory, despite the fact that in the former the state function varies with time and the operators do not and in the latter the reverse is the case.

A more controversial case from physics is the famous pair, curved spacetime vs. flat spacetime with compensating metric and force fields. Here some (Eddington, Schlick, Reichenbach) assure us that we have a classic case of theoretical equivalence, while others (at various times Putnam and Glymour, for example) think, rather, that the "theories" are not equivalent at all and that, indeed, the former is much preferable to the latter as a plausible account of the data. We will have more to say about this example later. Finally, we have the positivist philosophers' paradigm of equivalence, material world vs. malevolent demon or brain-in-a-vat, an alleged equivalence which, as a consequence of positivism, is taken by many (including Field, above) as a *reductio ad absurdum* of the positivist position.

The positivist position on theoretical equivalence, whatever one thinks of its adequacy or ultimate tenability, is at least clear. It is especially designed to handle those cases where we want to claim that surface incompatibility masks underlying commonality of import. The line is familiar. One discriminates among the consequences of the theory a special class of sentences which allegedly contain among them all the consequences of the theory open to empirical determination of truth independently of assuming the truth of the theory in question. If two theories agree on this observational set then they are alleged equivalent, no matter how radical the apparent incompatibility which resides in the remaining, theoretical, level of the two theories. The vices of the positivist program are well known, but here I would like to remind us of some of its virtues.

However fraught with untenable dualisms, incoherent foundationalist notions, etc., positivism may be, it at least offers an account of theoretical equivalence which brings into a harmonious whole a theory of meaning, of evidence, of ontology, of truth, of explanation, and of equivalence itself. Observationally equivalent theories do not differ in meaning since the total meaning of the theory is exhausted by its set of observational consequences. Since the consequences of the theory exhaust the totality of possible evidence for or against a theory, evidentially equivalent theories are automatically genuinely equivalent, blocking any opening for skepticism. One need not take the apparently conflicting ontological claims of the theories seriously, since the ontology we are committed to by apparently referring terms at the theoretical level is only *façon de parler* in any case. Nor need we be concerned about which theory is true, since, once again, all the theories really say is what their observational consequences say, and at that level they say the same thing. Finally, we are offered an account of explanation, basically the subsumption of particular correlated observable events under general rules of constant conjunction, which makes it clear that despite apparent differences theories equivalent in the positivist sense offer the same explanations of events, if explanation too is positivistically understood. I am, of course, laying all this out here grotesquely sketchily and without anything like the necessary attention to either detail or to variance and controversy within the positivist camp, but it will have to do for the time being.

Nor will I be concerned here with the manifold, familiar objections to the positivist program, replete with the well-known proofs of its alleged impossibility and absurdity. I will focus, rather, on one particular kind of realist alternative to positivism. In particular I will explore one way a realist might try to offer an alternative to the positivistic account of theoretical equivalence and will explore the problems encountered in such a program which an antirealist might put forward as reasons for skepticism regarding the viability of a notion of equivalence differing from his own. I make no claim that the realist account of equivalence I will offer is the only one a realist might offer, nor that this particular account of equivalence could not possibly overcome the difficulties I will raise in its path. I do think, however, that the account I offer does capture the instincts many realists have had regarding theoretical equivalence, and the problems I lay out do capture the skeptical doubts regarding such a realist program latent in many recalcitrant positivist's skeptical remarks.

What is wrong with the positivist notion of equivalence according to the realist? And how is the central deficiency of this account of equivalence to be remedied? What is wrong with the positivistic notion, according to the realist I have in mind, is this: Designed to handle plausible cases of theoretical equivalence, say the Schrödinger and Heisenberg formulations of quantum mechanics, the positivistic notion of equivalence misconstrues just what it is about these two theory presentations that makes them merely two expressions of one and the same theory. His misconstrual leads him, fallaciously, to believe that any reasonable commitment to the equivalence of theories must force one by an irresistible slippery-slope argument to such absurd conclusions as the equivalence of a realistic account of the world with an extended-dream account or a brain-in-a-vat account. The source of this misconstrual, says my realist, is the positivist's failure to take sufficient account of the interrelationships theories can bear to one another at the theoretical level. Once one sees that in all the plausible cases of theoretical equivalence there is a far stronger relationship between the two theory presentations than merely a common ability to save the phenomena, a relationship at the theoretical level, one will realize that it is this stronger relationship which warrants the justified claim of theoretical equivalence. But then we will realize that we are by no means forced from the acceptance of such genuine cases of equivalence into any assertions of equivalence of the kind devastating to the realist program.

So what I wish to explore is this: What is the structural relationship between theories sufficient for a realist to attribute equivalence to them? In particular, how does this additional requirement go beyond the weak positivist constraint of merely having all observational consequences in common? I will argue that the realist will find, unless he is very careful, that his program all too often seems to lead him into assuming far more in common with the positivist than he might at first think. And, I will argue, it is incumbent upon the realist to offer us not just a notion of equivalence more stringent than that of the positivist, but an integrated theory of meaning, truth, ontology, confirmation, and explanation into which his notion of equivalence will naturally fit. I certainly will not maintain that this is impossible for the realist, but only insist that profound difficulties must be faced up to by any realist who takes this program seriously enough to counter the positivist's claim that, dubious as his account may be, it at least offers us a fully integrated, coherent account of all these metatheoretical aspects of theories.

II

The realist I have in mind argues that it is not sufficient for theoretical equivalence that the two theories in question save all the same phenomena. A more profound and tighter structural interrelationship between the theories, an interrelationship which takes account of the relationship of the structure of one theory at the theoretical level to the structure of the other theory at its theoretical level, is required for genuine equivalence. What could this additional structural component be?

One thing we can be sure of. Whatever this structural "isomorphism" is to be, it cannot be a purely formal notion. It cannot be, that is, an interrelationship which can be determined to hold solely on the basis of the logical form of the theories in question. Why not? We will consider the strongest such possible formal interrelationship one can imagine. Let us suppose that the theories are term by term "intertranslatable," that is, that each can be obtained from the other merely by a substitution of terms of one theory for terms of the other. Would that be enough to show the theories equivalent?

Surely not. Let the two theories be 'All lions have stripes,' and 'All tigers have stripes,' with all the words in both theories taking on their usual meanings. The theories are intertranslatable in the purely formal sense. They are exactly alike in logical form and one can be obtained from the other by a simple term-for-term substitution. But they are most assuredly not equivalent. I am not denying that if we found a speaker who persistently asserted that lions had stripes and tigers didn't that we would probably take him to mean by 'tiger' what we mean by 'lion' and vice versa. Nor am I denying that it is just the question of the meaning of the theoretical terms of a theory that is at issue in many questions about the necessary and sufficient conditions for theoretical equivalence. We will return to that issue in detail. All I am claiming here is that mere commonality of logical form, even of a total theory when compared with another total theory, is certainly not by itself sufficient for theoretical equivalence. The meanings of the terms in the theories, however construed, are crucial to questions of equivalence.

But "translatability" of one theory to another, in some sense, is just the additional constraint, over and above saving the phenomena, that some realists want to demand as necessary for theoretical equivalence. So if that isn't the purely formal notion of commonality of logical form, what is it?

Two options suggest themselves. If we have some grasp of the meanings of the terms of the theories which comes to us from outside the role these terms play in the theories, and if on the basis of this knowledge of meanings we can affirm the logical equivalence of the theories, then, of course, the theories are equivalent. Straightforward translations of a theory phrased in one language to a version of the theory in some other language, where the theories are small fragments of the totality of beliefs of speakers of the languages and where the full vocabulary of the theories appears embedded in a far broader context than the particular theory in question and in such a way that we are inclined to say that we grasp the meanings of the terms independently of the role they play in the theories in question, are of this sort. Thus we won't have any trouble affirming that the theory that 'Salt is white' is equivalent to the theory that 'Salz ist weiss.' But this notion of equivalence will hardly be of help to us in the most interesting and crucial cases. Our conviction that Schrödinger quantum mechanics is equivalent to Heisenberg quantum mechanics hardly comes from some outside grasp of the meanings of the respective psi-functions which informs us that one simply means the time transform of the other.

So what does assure us over and above mere formal similarity, which we saw was never enough, that in this and similar cases the theories are genuinely equivalent to one another? I think the answer is clear. We believe that we have no antecedent understanding of the meanings of the theoretical terms (psi-functions and operators in the example of quantum mechanics) other than the role they play in the theories in question. We have no external grasp on the meanings of these terms that we could go to to determine whether the existence of a formal mapping from one theory to the other did or did not constitute a genuine translation, and hence a genuine demonstration of theoretical equivalence. Rather, all we have is this: Relative to whatever understanding we have of the notion of observational consequence of a theory, we are satisfied that the two theories in question really do have all their observational consequences in common. Over and above this we can demonstrate the existence of a formal mapping of some appropriate kind between the theories at the theoretical level. For the purpose of our argument it really won't matter very much just exactly what this mapping, necessary to show appropriate commonality of theoretical structure, amounts to. Together these *two* features of the theory, commonality of observational consequences and the existence of the appropriate structural mapping at the theoretical level, are taken to be enough to establish theoretical equivalence.

The intuition behind the position is clear. The meaning of the theoretical terms is fixed entirely by the role they play in their respective theories, in the holistic account of the meaning of theoretical terms now so familiar to us. The combined force of observational commonality and structural correspondence at the theoretical level is taken to be sufficient to establish commonality of meaning for the theoretical terms, their having, as they do, their meaning entirely determined by the place they occupy in the theoretical structure which generates the observational consequences. The details of the argument will depend on one's detailed account of the holistic structure of theories and on what one takes to be a sufficient structural interrelationship for theoretical equivalence. But consideration of such simple cases as theories which have common observational consequences and which can be obtained from one another by such simple operations as interchanging of terms for terms ("You just are using 'positive charge' to mean what we meant by 'negative charge' and vice versa") will be enough to give us the general idea of the motivations and arguments of the program.

Seen from this point of view a realist conception of theoretical equivalence should not be viewed as a total rejection of the positivist account. Rather, it amounts to the claim that the positivist constraint of commonality in saving the phenomena, while necessary for theoretical equivalence, is simply not sufficient. Again the intuition is clear. Two theories might have all the same observational predictions but be so radically different in their structure at the theoretical level that one ought to take them as attributing (realistically) quite different explanatory structures to the world. Only commonality of structure at the level of the theoretical ontology introduced by the theories to explain the commonly predicted observational results is enough for us to say that the two accounts are genuinely, realistically, equivalent to one another.

Notice how much of the positivist account the realist, construed this way, must first accept. The project of characterizing a notion of theoretical equivalence in this way at least presupposes that the notion of commonality of observational consequences is a coherent one. The argument which takes Schrödinger and Heisenberg to be offering us the same theory must at least assume that some coherent limits to the notion of observationality can be given. If the psi-function itself could be considered, under any circumstances, an observational quantity, then the existence of the appropriate mathematical transformation from one representation to another would no more constitute a demonstration of equivalence than does the trivial formal transformation from 'Lions have stripes' to 'Tigers have stripes' demonstrate the genuine equivalence of those assertions. Of course the realist need not lay down where observationality ends and theoreticity begins. He need only be assured that the consequences common to the two theories (in the example chosen probabilities of outcomes of measurements in the quantum mechanical sense) *include* anything which could be called observational. He must, that is, be sure that the apparent incompatibilities between the theories, incompatibilities he will demonstrate to be only apparent by means of his equivalence-establishing interrelationship shown to hold among them, are all firmly "trapped" at the nonobservational level. At least that much of the presuppositions of positivism is presupposed by this kind of realist as well.

To summarize, what my realist asserts is this: The point of theories is to introduce theoretically posited structures to explain the observable phenomena. Merely predicting the same phenomena is not enough for two theories to be equivalent, for they may explain these phenomena in radically different ways. But if the theories share both commonality of predicted phenomena *and* an, appropriately characterized, commonality of structure at the theoretical level, as demonstrated by the existence of the appropriate structural mapping between them at the theoretical level, then the theories plainly are equivalent. Any residual appearance of incompatibility must be due to merely verbal equivocation at the theoretical level. But realizing as we do, on the basis of the theory of the meaning of theoretical terms which tells us that the totality of their meaning is the place they play in the theoretical structure in question, that the apparent disagreement between the two theories is a mere superficiality of alternative verbal designations for common structural elements, we will not be deceived but will realize that, properly speaking, we have only one theory expressed in two misleadingly different ways. But it is essential for genuine equivalence that two conditions be met. On the observational level the theories must be identical. On the theoretical level they must bear the appropriate structural similarity.

III

What interrelationship at the theoretical level should the realist demand, over and above commonality in saving the phenomena, before he admits two theories to be equivalent? I hope I don't have to say, for each specific proposal, be it term-by-term translation, common definitional extensions, or whatever, would require its own careful analysis. I hope that I will be able to say all I wish to say without pinning the imagined realist or myself down to a specific proposal. But then how can we say anything interesting about potential pitfalls in his realist path?

One obvious problem for the realist is getting his principles to coincide with his intuitions. He may, in the so-familiar manner, find that his proposal is too weak or too strong in its demands to coincide with what he takes intuitively to be genuine cases of equivalent theories and what he takes to be cases of nonequivalent theories which happen to save the same phenomena.

Reichenbach, for example, wanted to hold, along with Eddington and Schlick, that curved spacetime and flat spacetime with universal forces were equivalent theories. But on a model of equivalence suggested by *some* of his remarks, remarks to the effect that equivalent theories simply called the same theoretical entities by alternative names, the model of equivalence he should have in mind would require a strict term-by-term interdefinability of the two theories. This won't be the case, however, since the flat spacetime theory, being "ontologically otiose" compared to the curved spacetime theory, can't have its parameters defined from those of the ontologically more parsimonious theory. That is, given a flat spacetime metric and the "universal forces" we can determine spacetime curvature in the curved spacetime theory. But a full specification of curvature is not enough to *uniquely* specify the proper flat spacetime and universal forces. Crudely, this is related to the nineteenth-century observation that uniform gravitational fields are empirically undetectable. (I will have more to say about this case later.) So one will either have to give up his stringent model of equivalence or renege on his intuition that these two theories are, indeed, equivalent. In the case of Reichenbach matters are more complex, of course. Actually, what he does is to suggest in other, dominating places that commonality of observational consequences is sufficient for equivalence, which certainly saves the equivalence of curved spacetime and flat spacetime plus universal forces. Naturally, though, adopting this notion of equivalence plays havoc with his alleged scientific realism.[2]

One can see this problem of a conflict of intuition with principles arise in the other direction as well. Even quite stringent realist notions of equivalence may not be strong enough to exclude as equivalent pairs of theories the realist does not wish to think of as equivalent.

Surely the realist, although he wants Schrödinger and Heisenberg to say the same thing, does not want the usual wild Cartesian possibilities to count as theories equivalent to our ordinary scheme of the external material world. Now some Cartesian alternatives, say the one that tells us that there is simply nothing but our private experience and nothing "out there" to explain it, will be rejected as inequivalent to our ordinary theory because of a lack of structural similarity to our ordinary theory on the theoretical level. But other Cartesian fantasies, say the one in which all our private experience is caused by an appropriate signal fed into our tank-immersed brain along a cord (or caused by the appropriate state of the cybernetic machine at the other end of the cord), look as though they have the possibility of being made to look as structurally similar to our ordinary "external world" theory as we like. Indeed, what seems to differentiate the real cases of discovered scientific equivalence from these spurious Cartesian cases is not interdefinability in the former cases and not in the latter. Rather, what makes the scientific cases interesting scientifically and the Cartesian cases not is that it is a significant scientific task to demonstrate interdefinability in the scientific cases, whereas interdefinability is just too trivial to be interesting in the kind of Cartesian cases I have in mind.

2. Reichenbach's allegation of the equivalence of curved spacetime with flat spacetime plus universal forces can be found in H. Reichenbach, *The Philosophy of Space and Time* (New York: Dover, 1958), secs. 3–8. The contrasting "realist" attitude toward theories is in his *Experience and Prediction* (Chicago: University of Chicago Press, 1938). Specifically useful on the problem in Reichenbach, and generally useful as background to this essay throughout is C. Glymour, "Theoretical Realism and Theoretical Equivalence," *Boston Studies in the Philsophy of Science* 8 (1971): 275–88. See also W. Quine, "On Empirically Equivalent Systems of the World," *Erkenntnis* 9 (1975): 313–28.

Here quite a bit of caution is called for, though. There is the kind of realist who rejects the holistic-role-in-a-theory account of the meaning of theoretical terms. For him there is no problem in asserting that the Cartesian alternatives, no matter how structurally similar to our ordinary external-world account, are not equivalent to the ordinary account. But then just what account of theoretical equivalence he will offer, at least what account which will handle the cases we are interested in where we think equivalence is clear despite surface incompatibility, isn't clear. The realist we have in mind, espousing as he does the doctrine of meaning which gives to theoretical terms only the meaning they possess from their holistic role in the theories in question, will find it difficult, I think, to avoid attributing equivalence of the suitably constructed Cartesian alternatives to our ordinary account. But then his account of theoretical equivalence is sliding perilously close to the positivist account. For if the brain-in-a-vat account of the world is really equivalent to the ordinary material-object world account, so long as the brain-in-a-vat account is suitably formally structured, then have we really gotten very far from merely saving the phenomena as sufficient for theoretical equivalence? Once again the tension between an "instrumentalistic" account of the meaning of theoretical terms and a realist account of theoretical ontology is clear. But our realist can just bite the bullet and affirm that, appearances to the contrary, such cleverly tidied-up Cartesian alternatives are simply equivalent to our ordinary world-view. Contrary to the view expressed in the quote of Field at the beginning of this paper, all "evil demon" hypotheses would not then be equal. Some would be genuine alternatives to our ordinary world-view, presumably to be rejected as unacceptable on some epistemic ground or other. Other "evil demon" hypotheses would just be our ordinary world-view dressed out in peculiarly misleading terminology. It all depends on whether or not our hypothesis about the doings of the evil demon has him failing or succeeding in structurally duplicating our ordinary world in his alternative causal structure of our private experiences.

IV

Let us suppose that our realist is satisfied that his formal notion of theoretical equivalence is consonant with his intuitions about just what pairs are pairs of genuinely equivalent theories. What further concerns must then arise?

One is fairly obvious given a principle component of the motivation behind the positivist position on equivalence to which our realist is opposed. Surely one of the major thrusts behind the positivistic notion is its ability to limit the possibilities of skepticism. Faced with alternative theories all equally compatible with all possible empirical data, we are at a loss as to how to decide which of these we ought to believe. Taking the positivist line and adjudicating all of these theories as saying the same thing, there is no longer a decision to be made. Hence there is less room for skepticism to intrude. We still are faced with the inference problem from observed empirical phenomena to full generalizations over all possible empirical phenomena, of course the problem of induction in its most general form, but at least we need no longer fret about alternatives at the level of the in-principle unobservable haunting us.

Once we insist on stricter conditions for equivalence than mere commonality at the empirical level, the problem of rationalizing the choice of alternatives to be believed is reinstated. Nor, from a realistic point of view, with its insistence on truth as the goal of inquiry and on a correspondence theory of truth (whatever that means) even at the theoretical level, do such ways out as allowing rationality to be permissive ("You are rational if you believe any one of the nonequivalent theories which saves the phenomena equally well") appeal. What we have a right to expect from the realist is a systematic account of what rationalizes the choice of one theory as true over any of the nonequivalent but empirically indiscriminable alternatives to it.

The realist can, of course, simply refuse the challenge and accept the skeptical consequences. But suppose he doesn't. What we will be offered is an account of the rationality of theory choice which will allow us to go beyond the conformity of a theory with all possible empirical data in deciding its worthiness for our belief. One problem will be, though, that we will need to be assured that the powers of this model of confirmation are sufficient to bring us up to the boundaries of equivalence. To avoid the intrusion of skepticism, that is, we will need assurance that for any two nonequivalent theories one will be preferred epistemically to the other. In other words there will have to be a close internal harmony between the notion of theoretical equivalence offered and the notion of the epistemic worthiness of theories.

Without a specific model of theoretical equivalence and a specific, allegedly harmonized, model of confirmation, I can only illustrate this problem by briefly considering in a qualitative way the issues I have in mind. Typically one can generate alternative theories saving the same phenomena by some process which introduces into a theory otiose elements whose place in the theory "cancels out." Most interesting, of course, are the historical cases where the theory with the otiose elements came first and where it was an important scientific discovery that one could eliminate them by a conceptual revision. For example, one could replace the aether theories used to account for the null round-trip experiment results with special relativity, thereby replacing a manifold of theories with a "cancelled out" element (the velocity of the observer relative to the aether) with a theory without that vacuous element. Again Einstein showed us that one could replace earlier gravitational theories with their otiose elements of "real inertial frames" and "real gravitational fields" by curved spacetime with its merely local inertial frames.

While the positivist seems to be committed to the view that a theory and its excessively otiose counterparts are all equivalent, the realist will, in general, claim that they are not. What we can expect from him is some attempt at a systematization of the notion of ontological simplicity which will credit the less otiose theory with greater simplicity and the simpler theory with greater warrant for belief. Obviously this will require a notion of confirmation which differs from that which associates confirmation with having the right observational consequences, and attempts in this direction have been made.[3]

But what if we have a pair of theories such that each member of the pair is preferable to the other in ontological simplicity in some respect? We can easily imagine a situation where A is preferable to B in terms of ontological parsimony for one aspect of theoretical structure, and B preferred to A along another component. Here one will probably hope to establish that relative to this pair of theories there is some third alternative, C, which is ontologically more parsimonious than either A or B, and, hence, preferable to both of them. What we will expect from the realist, though, if he takes this approach, is some reason to believe that relative to the notion of

---

3. For an attempt at a theory of confirmation appropriate to this realist notion of equivalence, see C. Glymour, *Theory and Evidence* (Princeton: Princeton University Press, 1980), especially chaps. 5 and 9.

theoretical equivalence which he espouses, in each possible case there will always result a single unique "maximally parsimonious" theory. Or, rather, that relative to his notions of theoretical equivalence and ontological parsimony, any two theories which, relative to a given set of observational consequences, are both adjudged maximally parsimonious will turn out, on his criterion, to be theoretically equivalent. Unless he can show us that there is a most parsimonious theory, and that this is unique modulo his notion of theoretical equivalence, we will once again be faced with room for the intrusion of skepticism.

But there is a more profound problem of an epistemic sort than this for the realist. Taking equivalence to demand more than commonality of observational consequences the realist is faced with the threat of skepticism which arises when he tolerates inequivalent theories having all their observational consequences in common. So his confirmation theory is supplemented with elements going beyond reflection on the observational consequences of a theory in evaluating its merit for belief. Whether this be ontological simplicity, as I have chosen for the example above, or conservatism with respect to antecedent theory which has frequently been proposed, or some other possible feature of theories altogether, is immaterial. The first problem the realist faces, unless he is simply willing to tolerate states of irresolvable suspension of belief, is to try to demonstrate to us that his combination of a notion of theoretical equivalence and a notion of warranted believability always produces as most believable only a single theory or a class of theories all equivalent to one another relative to his criterion of theoretical equivalence.

But even if he does this should we then be fully satisfied? I think not. For the question will always arise: Why should we accept *this* notion of confirmation or epistemic preferability? Why should we believe that the simpler theory, or the theory which deviates minimally from previously accepted theory, is the theory which is, in any sense, most likely to be true? The point is hardly new. Insofar as one places in one's structure for warranted belief any considerations that go beyond mere conformity with the observational data, it is hard to see why one ought to take these additional espoused "marks of believability" as genuine "marks of truth."

Of course there are the familiar options open of adopting as a priori principles that simpler theories are more likely to be true, etc.

I do not wish to pursue these here. We ought to note that one familiar approach is not open to the realist we have in mind. There is a familiar "justification" for such principles of warranted belief of a pragmatist sort which claims that the normative principles of believability rest ultimately on our actual practice of belief, and that the connection between truth and believability rests not on showing that some process for belief leads (more likely than not) to the truth, but rather on understanding the very notion of truth as ideal warranted believability. However persuasive such a tack may be in trying to rationalize our actual reliance on principles of simplicity, conservatism, etc., in science, the realist we have in mind should be loath to undercut the possibility of skepticism by this means. For if there is anything realistic to his realism, it should include at least the rejection of such a pragmatist line on truth and on justification. Whatever realism about theories means, it should include a claim that truth is correspondence to objective reality (in the metaphysical sense in which this is a controversial thesis), and this brings with it strong pressure for an account of just *why* we ought to accept anything over and above conformity with the observational data as relevant to the believability of a theory. But some such invocation of principles over and above conformity with observational data is necessary for anyone who wishes to impose a constraint on equivalence which outruns commonality in saving the phenomena and who does not wish to accede to the possible skeptical consequences of that demand.

There is, of course, a familiar response the realist can give. Unfortunately, it is of the "You're no better" form with the usual unsatisfactoriness of the *tu quoque* reply. This response is the familiar one of the realist to the positivist that even eschewing postulation of unobservables still leaves us with the problem of leaping beyond present data to a fully general theory of the world. And, the reply will go on, the problem of purely inductive inference is as infected with the invocation of notions like "simplest hypothesis" etc., as is the discussion of inference from the observable to the unobservable in principle. Perhaps so. Still we shall want some rationalization from the realist for his invocation of whatever principles he chooses to suggest limit the range of skepticism inevitably introduced by his belief in an ontology of in-principle unobservables, and by his invocation of a criterion of theoretical equivalence which demands that in at least some cases theories be inequivalent even when they save all the same phenomena in all possible worlds.

V

If the kind of realist notion of theoretical equivalence we have been discussing runs into problems in the upward direction of confirmation, there are difficulties it must deal with in the other, downward, relation of theory to data as well. We take it that our theories are explanatory. How must the realist's notions of explanation and of theoretical equivalence mesh?

First there will be the obvious requirement that his account of explanation satisfy at least some version of an equivalence principle. If two theories are genuinely equivalent, then, one hopes, in at least some sense they ought to offer the "same explanation" of the phenomena. And if the theories are inequivalent on the realist's criterion, then there ought to be a discrimination on the level of explanation which tells us why they are not, even if they are predictively equivalent, explaining the predicted phenomena in the same way. Actually the case here is subtle, for there might be alleged to be a kind of intensionality which allows theories to be equivalent in the sense the realist has in mind yet not fully equivalent in some "finer grained" explanatory sense. The individuation of "kind of explanation" might, for example, depend upon alleged modes of expression so that one and the same theory might in two equivalent versions be said in some sense to offer different explanations due solely to the mere difference in the manner in which the theory is expressed in the two versions. Someone who finds the notion of explanation loaded with "pragmatic" aspects might indeed hold to this. In the same vein it would, I suppose, be possible to maintain that two equivalent versions of a theory received differential confirmation from the same data, despite the obvious persuasiveness of a principle of equivalence as a criterion of adequacy of an acceptable notion of confirmation.

What is certain, I think is that an exponent of a realistic notion of equivalence must needs be quite wary of allowing equivalent theories to either offer inequivalent explanations of the data or to be inequivalently confirmed by the same evidential base. If he allows both of these, his notion of equivalence is beginning to drain into a bloodless ghost of that notion as preanalytically understood. At the very least we would expect from him coarse-grained notions of explanation and confirmation which would mesh with his notion of theoretical equivalence in the natural way, and an account as to why the finer-grained notions, which allowed equivalent theories to offer inequivalent explanations or to be inequivalently confirmed, did not vitiate the real sense of theoretical equivalence he originally had in mind.

Once again, a detailed critique of a realistic attempt to "mesh" notions of theoretical equivalence and explanation will be impossible for us here, for that would require what I am trying to avoid, the presentation of a specific principle of equivalence and specific accounts of explanation and confirmation. Instead I will focus on what to the positivist seem to be certain fundamental difficulties with any such realistic program. What the positivist will allege is familiar. Suppose we have available that minimal, best confirmed theory compatible with the data (and, of course, all of those equivalent members of the equivalence class of theories of which this particular one is representative). Presumably this theory is the best possible explanatory account of the data. Indeed, any reasonable notion of confirmation and explanation should lead us in at least some weak sense to take it that the best confirmed, most plausible account is the best explanatory account, a relationship which becomes, of course, trivial if we adopt notions of confirmation as inference to the best explanation.

One thing had better be the case if the realist account is to hold onto even a skeletal version of its fundamentals. That theory which is the Craigian reduction of the theory in question had better not appear among the set of equivalent best alternative theories. For if it does the notion of theoretical equivalence in mind would seem automatically to reduce to the positivist notion. But how is the realist to avoid the introduction of the Craigian reduction, if not as equivalent to his best theoretical account, then as a simpler (hence, superior) alternative account to it?

Relying on the syntactical complexity (or, indeed, infinitistic nature) of the Craigian alternative to exclude it seems rather "unrealistic" in nature, a move more suitable to a frank pragmatist notion of theoretical preferability. Surely the right way out for the realist is to argue that the Craigian alternative is to be rejected for its explanatory inadequacy. According to the realist, the best realistic theory and its equivalent alternatives explain the observational phenomena in a way in which the Craigian reduction does not. Indeed, the Craigian surrogate is alleged to be devoid of explanatory power at all. Theoretical postulation, we are told, does not merely summarize observational generality nor accommodate it in a compact syntactical form. Theoretical postulation is taken to *explain* the observational generality. Hence the mere statement of that generality is, in a fundamental sense, totally devoid of any real explanatory power.

But what is this notion of explanation which makes the best realist theory explanatory in a way the Craigian reduction of it is not? Here we must once again focus on two quite distinct kinds of realism, one the kind of realism which takes a view of the meaning of theoretical terms as given by their holistic role in the theories in which they appear, and the other which allows the attribution of "excess" meaning to theoretical terms, by semantic analogy or otherwise. Thus, for example, for the first kind of realist molecules exist, but the meaning of 'molecule' is given solely by the complex role it plays in the sum total of molecular theories. For the other kind of realist, we understand 'molecule' by means of our understanding of such notions as 'particle,' notions whose meaning is originally accrued in their use to refer to observables, but whose meaning is preserved intact when the term is used in the quite different contexts of reference to "particles too small to see" or even "particles unobservable in principle." The first kind of realist, the one with the idea of the meaning of theoretical terms as being solely their role in the holistic theory, is the one we have been emphasizing. Why? Because, I have claimed, it is only for such a realist that the notion of theoretical equivalence as being structural isomorphism super-added to observational equivalence is plausible. Without that theory of the meaning of theoretical terms, it is hard to see why even the most strict commonality of structure on the theoretical level would be enough to guarantee equivalence. But, I want to argue here, it is from the point of view of the second kind of realist that the notion of realistic explanation as something over and above subsumption of observed facts into generalizations about observables makes its most plausible sense. In other words, I want to argue that there is a fundamental tension between the realist's desire to posit a notion of explanation over and above that adopted by the positivist, and his desire to maintain a theory of the meaning of theoretical terms which allows for the notion of theoretical equivalence we have been discussing.[4]

Suppose we have a generalization over observable phenomena. What does the realist demand over and above that? Presumably the postulation of a theoretical entity or property which explains the observational generality. But what additional explanatory power does this theoretical postulation give us? A frequent answer is "unification." Positing the theoretical structure provides, somehow, a *unified* account of the phenomena.[5]

Let us look at some cases to see why I think there are difficulties for the realist here. The observational facts of dynamics require us to single out a set of preferred states of motion, the local inertial motions. Additionally, optical phenomena pick out as distinguished these same reference frames. The realist accounts for all of this by positing the spacetime structure itself as existing over and above the (for this purpose) observational moving systems and light waves. The dynamic and optical distinctions between inertial and noninertial states are then accounted for by the relation of the physical systems to the underlying spacetime structure. But, replies the positivist, what kind of additional explanatory force does positing "spacetime itself" give us? Whereas before we had the irreducibly inexplicable basic distinction between inertial systems and noninertial, now we have a new, and trivially introduced, "deeper" explanation in terms of the underlying spacetime and the relations of the physical systems to it. But have we done anything more than added an otiose layer to what was already a perfectly adequate theory, a layer which only gives the spurious appearance of further "unification" or further "explanation" to the phenomena?

Many positivists would argue like this: The appearance of additional explanatory value in the substantivalist spacetime account, over the purely relationist theory, is due to the fact that while the reference to spacetime itself is really introduced only in the holistic-place-in-the-theory manner of the realist we have been empha-

---

4. For a discussion of that notion of the meaning of theoretical terms which allows meaning to accrue to them over and above the role they play in the theory, see the author's "Semantic Analogy," chap. 7 in this volume.

5. On realistic theories as explaining by unifying, see M. Friedman, "Explanation and Scientific Understanding," *Journal of Philosophy* 71 (1974): 5–19, and his *Foundations of Space-Time Theories* (Princeton: Princeton University Press, 1983), especially chap. 6.

sizing, the naive picture of spacetime as a kind of "ghostly" substance, sort of like a thin "rigid" extended material thing, gives us the impression that we are getting an explanation like that which the other kind of realist would offer. If we really understood what we were talking about when we posit the theoretical apparatus, independently of positing its explanatory role in the situation in question, then we could attribute further explanatory force to the substantival theory over the relational in the manner familiar from some older realists (reducing the unfamiliar to the familiar, demonstrating "mechanism" in the Newtonian sense, etc.) But we have no such grasp of what the theoretical apparatus is over and above its place in the theory in question. As a consequence, its putative explanatory force is void.

To see this, the positivist continues, consider another case. Study of the symmetries of the interactions of elementary particles gives us a systematization of these particles in terms of various symmetry groups. One can capture this symmetry structure by positing the existence of various "charges" for the particles (strangeness, charm, etc.) and associating with the posited charges various appropriate conservation rules. But in positing these charges and their conservation, are we *explaining* the symmetries involved? Most theoretical physicists would argue, I think rightly, that the invocation of the charges is simply another way of stating what the various symmetries are. There is a redundancy of the "theoretical structure" on the "observational data" to be explained. If anything would count as genuinely explanatory of the symmetries, they would argue, it would be the subsumption of these symmetries into some higher, more general symmetries naturally generated as the consequence of a more profound theory (say the generation of the symmetries out of some posited gauge field or other unified field account). The congruence of all of this with the positivistic notions of explanation is clear.

Now consider the alleged explanation of the symmetries of baryons and mesons given us by quark theory. Here we do seem to feel that a genuine explanatory account has been offered. Part of this intuition is, I think, perfectly acceptable positivistically, quark theory being more general and profound than the symmetry accounts derivable from it. But part of the intuition rests on our viewing quarks "analogically" as tiny particles constituting larger particles in a manner understandable to us from the whole-part relationship and its explanatory value in the realm of observables.

Once again we see that the invocation of terms whose meaning rests solely on the role they play in the appropriate theory suggests an account of explanation which is wholly positivistic, whereas the other kind of realism, the one which allows terms to have "excess meaning" attributable to them through semantic analogy seems to fit more harmoniously with the realistic notion that explanation by positing of theoretical structure is more than the subsumption of observational facts under broad generalities.

I do not, naturally, intend to pursue this problem of explanation from the realist point of view in any depth here. To repeat, my point is only this: A realist may take two attitudes toward the meaning of theoretical terms, either the holistic role-in-the-theory attitude or the alternative which grants to theoretical terms meaning accrual over and above the role the terms play in the theory. From the latter point of view it is fairly clear how we will be offered a realist account of explanation which takes explanation to be something over and above hierarchical subsumption of observational possibilities into generalities. But, from this point of view, it is hard to see what the realist account of equivalence of theories will amount to. Basically it must come down to the line that theories are equivalent only when they "say the same thing," and this notion will have to be explicated in terms of whatever theory of meaning (over and above role-in-theory) the realist offers us. From the alternative point of view regarding theoretical meaning a much simpler notion of theoretical equivalence can be constructed. This is the notion which simply takes the positivistic notion of equivalence as necessary and super-adds to it some notion of structural isomorphism between the theories at the theoretical level to obtain a sufficient condition for equivalence. But his notion of theoretical meaning, and of theoretical equivalence, besides borrowing many dubious presuppositions from the positivist position, suffers also from the difficulty of making it hard for us to see just what it is that is explanatory in a theory, or in any of its equivalents, which is not already there in the allegedly nonequivalent and nonexplanatory Craigian reduction of the theory so beloved of positivists.

VI

To summarize: Positivism, for all its defects, offers us a theory of theoretical equivalence neatly integrated with its theory of confirmation and its theory of explanation. The realist is obliged to do the same. If the realist adopts a theory of the meaning of theoretical terms which attributes to these terms a meaning over and above that which accrues to them by means of the role they play in the theory, then his notion of theoretical equivalence will be complex and will depend in detail on just what theory of meaning for theoretical terms he offers us. If he adopts a theory of meaning for theoretical terms which takes their meaning to be fixed solely by the role they play in the theory, then his notion of theoretical equivalence is likely to be simpler. It will be the notion of two theories, first of all, sharing commonality of observational prediction and, secondly, having an appropriate structural isomorphism at the theoretical level.

One will first ask if the notion of equivalence offered corresponds with the realist's intuitions. Does it take as equivalent theories only those the realist wishes to count as equivalent and will it count as equivalent all those pairs preanalytically thought equivalent?

Next, one will want to examine the interaction of the notion of equivalence offered with the "upward" notion of confirmation. Will equivalent theories always be equally confirmed by the same data or, if not, will the violation of the "equivalence condition" for confirmation receive an adequate explanation? Will the theory of confirmation allow for differential confirmation of theories designated nonequivalent, or instead, will it leave room for skepticism. Even if the theory of confirmation does always select from a group of nonequivalent alternatives a unique "best confirmed" member, will there be an appropriate *realist* rationale for adopting that notion of confirmation?

Finally, will one want to look at the interaction of the notion of equivalence with the "downward" notion of explanation? Will equivalent theories always offer "the same explanation" of the phenomena and nonequivalent theories "different" explanations? More important, what notion of explanation, over and above the positivist notion, does the realist have in mind? Will it be such as to allow us to understand why a theory is not, according to the realist, equivalent to its Craigian reduction and why the theory with its realistic posits is explanatory in a way in which the Craigian surrogate is not?

Far be it from me to claim that these questions cannot be answered by the realist. But until they are we should be reluctant to dismiss the positivist notions with a sneering reference to "out-

moded verificationism." It is sometimes said (wrongly I believe)
that scientists do not reject one theory until a better one is available
to take its place. Just exactly what is the realist notion of equivalence, and which associated realist notions of confirmation and
explanation are supposed to take the place of the positivist's integrated if implausible accounts?

3. Facts, Conventions, and Assumptions in the Theory of Spacetime

This chapter is much the longest in this volume. It is also an essay whose organization is somewhat complex, and which deals with a number of distinct but rather intricately interrelated topics. For these reasons a somewhat extended guide to its argument is in order.

The basic theme of the essay is an argument to the effect that the formalization of a theory is not generally an empty exercise in logical sophistication, but is, rather, reflective of, and indicative of, numerous methodological and epistemological aspects of the context in which theorization takes place. In formalizing a theory some terms are taken as primitive and some are defined. Some propositions are taken as axiomatic, some as definitional, and some as theorems to be derived from the more fundamental axioms and definitions. While there is a certain degree of arbitrariness in choosing which terms are to be primitive and which propositions are to be basic, more often than not the choices made in a given formalization are not mere matters of convenience or logical elegance, but, instead, are reflective of an implicit metaphysical-epistemological critique of the theory being formalized.

To exemplify this claim this essay focuses on a number of test cases. All are drawn from recent theories of spacetime (special and general relativity) and all are concerned with formalizations of the theory which emphasize various ways in which the spatiotemporal concepts of the theory are to be interrelated to causal concepts and to various concepts appropriate to the description of idealized measuring instruments.

Are various spatiotemporal relations like that of simultaneity for distant events, or of the spatial and temporal intervals between events merely a matter of convention or decision on our part? Some philosophers have been prone to argue that if the world is as described by the special theory of relativity these spatiotemporal features are merely conventional in their

I am grateful to David Malament, John Stachel, and Robert Geroch for their advice on points of mathematics and physics.

nature. And, they have sometimes argued, this follows from the fact that if special relativity is correct, then there is no way to define *these spatiotemporal features of the world in terms of causal features.*

But is this so? John Winnie, expounding on somewhat neglected work by Robb, argued that since Robb has shown that these spatiotemporal relations were, *as a matter of fact, causally definable in a special relativistic world, they were not, contrary to the familiar claim, merely conventional. Here I argue that those on both sides of the debate have been too quick to leap to their conclusions. Very fundamental issues concerning what is meant by definability, what is meant by a causal relationship, what does and does not serve to support a claim of conventionality are all rather neglected in much of the debate. Only by bringing these fundamental philosophical questions to the surface, and by examining the multiplicity of answers one might give to them, can we fully understand how a result in mathematical physics like that of Robb can be understood to have its philosophical relevance.*

Robb takes a certain relationship among events as primitive, essentially the relationship which holds when one event is causally connectible to another and later than the other in time. He then gives us a number of axiomatic propositions we are to take as true of this 'afterness' relationship, propositions which are true of this relationship if special relativity is correct. He then offers definitions, in terms of the 'after' relationship of such notions as simultaneity for distant events, temporal and spatial separation, etc. If special relativity is correct, the relations so defined will coincide with the relations as more usually characterized in terms of measurements with ideal clocks, transported rigid rods, etc.

But if special relativity is not correct, if, rather, the spatiotemporal world is as general relativity describes it, then the congruence between relations as Robb defines them and as clock and rod defined will in general not hold. This may be the case even if the axioms governing the 'after' relations which Robb proposes still hold in the general relativistic world. And there will be other general relativistic worlds in which the axioms themselves are false, and the Robbian program therefore entirely vitiated. I explore in some detail what our reactions might be to the situation where Robb's axioms remain valid, but where the congruence he imagines between the relations defined his way and as defined the more usual way in terms of clocks and rods breaks down. I argue that the many alternative ways of viewing this situation which one can describe reflect profound differences of opinion one might have regarding such issues as what is to be taken as epistemically primitive in science, what constitutes meaning relationships among scientific terms, how one ought to regard the relationship of the theoretical to the observable in science, and what the role might be of the imagined possible hypotheses which form the background to one's scientific decision-making. I argue that even if special relativity were correct, the Robbian axioms and congruities holding, reflection on the variety of responses one would make to the theory suffering stresses in the way I describe is importantly relevant to our understanding of the philosophical consquences to be drawn from Robb's mathematical-physical results.

Next I move on to discuss, in a more general philosophical vein, various issues concerning definability of theoretical terms and the issue of the conventionality of theoretical beliefs. I place some of these issues in the general context of the alleged underdetermination of theory by all possible empirical data, and offer some reflections on how these general issues become particularized in this special case. Some further relevant consider- ations are then elaborated about the very point of so-called causal theories of spatiotemporal relations. The moral of this lengthy treatment of a spe- cific case is that any simple claim to the effect that a spatiotemporal feature of the world is conventional because it is not "causally definable" (or nonconventional because it is "causally definable") ought to be looked at with a good deal of skepticism. Fundamental issues of what definability is, what constitutes a causal relation, and what is contained in a claim of conventionality for a portion of physical theory are all fraught with subtly profound and difficult philosophical issues.

The remainder of the chapter is devoted to showing that the philosophi- cal issues raised in considering Robb's formulation of special relativity and its alleged consequences arise in other contexts as well. In particular, I examine some alternative proposals for formalizing metric notions in general relativity, and some proposals for associating, not metric, but topological aspects of general relativistic spacetimes with certain causal features of the world.

While the special theory of relativity allows but one spacetime, with a fixed metric, the general theory of relativity contemplates a variety of metrically different spacetimes as genuine physical possibilities. But, as in special relativity, the metric features of the spacetime are connected by the theory with a variety of other features of the world closer to the observa- tional level. Thus the behavior of ideal clocks, of ideal rigid rods, of light rays, and of particles not acted upon by forces (other than gravity) are all lawlike related to the metric of the spacetime.

Consequently, anyone wishing to determine the actual metric structure of the spacetime has a variety of procedures available to him to do so. Which procedures are best taken as defining the spacetime metric? Here I contrast the familiar procedures using rods and clocks with Synge's pro- posal, which uses clocks alone, and most important, with a formalization of general relativity, described by Ehlers, Pirani, and Schild, which uses only light rays and the paths of freely moving particles to fully map out the spacetime metric. I argue that some virtues alleged to this latter method are not as persuasive in its favor as might first be thought, and that, just as in the case of special relativity, the question as to which "definition" of the metric is most appropriate is one which requires both philosophical under- standing on our part of what a definition is supposed to do, and a grasp of the background realm of alternative theories which may someday be of interest to us as physical possibilities.

The earlier parts of this chapter cast doubt on the possibility of anything one might call a causal definition of the metric of spacetime being viable in the full general relativistic context. But the topological structure of a space- time is much weaker than its full metric structure. Could this portion of the spacetime structure be, in any plausible sense, causally defined? Here, once again, the issues are elucidated by comparing those situations where certain causal relations agree with the topological in their extension with those situations where they do not. Subtleties in the notions of what is appropriately called a causal relation, and in the notions of definition allegedly connecting topological to causal notions, are briefly examined, especially with relevance to the question of the very point of alleging that the spatiotemporal topological notion could be causally defined. The issues discussed here are taken up at greater length in chapter 9.

I INTRODUCTION

GIVEN a physical theory, there are always those who will try to formalize it. At a minimum this entails an attempt to present the theory informally in terms of a number of basic propositions about the world from which all the consequences of the theory in question can be derived informally. At a maximum such an under- taking involves an exact and rigorous presentation of these funda- mental propositions in some formal language (first order quantifi- cation theory or set theory, say), with the implicit claim that all the consequences of the original theory could, if they themselves were suitably "formalized," be derived formally from these rigorous fundamental propositions by the mere application of the formally specified rules of the logic in question.

Clearly such "axiomatization" can be a useful (or even crucial) component of ongoing science or the philosophy of science. Axi- omatization need not be merely a more or less careful display of some degree of technical ingenuity applied to no apparent purpose, even if that is what it frequently turns out to be. But just *how* axiomatization can be applied to scientific or philosophic purpose is not always completely clear from the axiomatizations themselves. The scope and limits of formalization as an aid to science, and especially to philosophy, are the subject of this essay. But I shall treat these general questions within the limited context of some formalizations of the theory of spacetime, since these particular cases provide highly illuminating examples of the general issues.

II FORMALIZATIONS AND THEIR DIFFERENCES

All standard axiomatizations have some familiar components. First of all there is the accepted logical framework, which may be restricted to "purely logical" terms or may encompass a vocabulary of mathematics that is in general taken for granted. Then there are the *primitive descriptive terms* of the theory in question. They are just "givens," and although they may be "explicated" informally outside of the formalization, nothing within the formal presentation pretends to "define" them or otherwise specify their "meaning." There are the *axioms,* statements accepted as true and containing in their vocabulary only the "logical" and primitive descriptive terms—at least initially.

Next there are the *definitions,* and over their status much philosohical debate can rage. In the definitions we have, standardly, new descriptive terms introduced on one side of an equation or equivalence, and phrases involving only the primitive vocabulary or previously defined terms on the other side. Not all equations can serve as definitions. They must meet the conditions of eliminability and noncreativity. If a definition is legitimate, one must be able to replace the defined term in whatever context it appears in further on in the formalization by some expression using primitive terms alone. The theory that results when the definition is included must not contain any consequences phrasable in terms of the primitive vocabulary alone that would not be consequences were the definition eliminated and the defined term replaced everywhere by its defining expression.[1]

Let us note some important ways in which formalizations may differ from one another:

(1) Two formalizations may have all the same consequences, but differ in that one and the same term appears as a primitive in

---

1. See, for example, P. Suppes, *Introduction to Logic* (Princeton: Van Nostrand, 1957), chap. 8.

one formalization and as a defined term in the other. Naturally this will force the formalizations to differ also in which consequences they count as definitions and which as postulates.

What we count as a primitive term and what as defined are in some sense not "internal" to the formalization but imposed upon it by us. That is, this differential attribution of status to terms is not forced upon us in any way by the consequences of the formalization. Let us call formalizations that differ only in regard to which terms are considered primitive and which defined, formalizations differing merely in the *status attributed to terms.*

(2) Two formalizations may agree as to which terms are primitive and which defined. And they may agree in that the consequences of each framable in primitive terms alone are the same. Yet they may still differ in at least one consequence which is phrased using a defined term. Let us say that such formalizations differ merely in *definitional consequence.*

(3) Suppose two formalizations differ in what they both agree to be a consequence framable in primitive terms. Then, as we shall see, all parties are likely to agree that the formalizations are based on different theories. But suppose two formalizations differ in some consequence. And suppose one formalization so categorizes terms as primitive and defined that it declares the differences to be only in definitional consequence. Suppose further that the other formalization declares some of the differential consequences to be framable in primitive terms only, given its classification of terms as primitive or defined. Let us call formalizations that differ in this way, formalizations differing only in *one-way definitional consequence.*

(4) Two formalizations may agree as to which terms are primitive and which defined; and they may have exactly the same consequences. But one formalization may declare, "in the margin," some of these consequences to be definitions that the other calls postulates or axioms, and vice versa. Let us say that two such formalizations differ only in the *attribution of definitional status to consequences.*

The relevance and importance of these distinctions will become clear as we proceed. I shall be concerned primarily with two fundamental questions:

(1) Given two formalizations that differ in one of the ways noted above, what good scientific and/or philosophical grounds could there be for preferring one such formalization to another?

(2) What are the scientific and/or philosophical consequences of opting for one such formalization over an alternative that differs from it in one of the ways noted above?

III FORMALIZATIONS AND MINKOWSKI SPACETIME

Robb's Formalization of Minkowski Spacetime

Instead of continuing the discussion at this abstract level, let us descend to a concrete case. I shall start with Robb's ingenious, elegant, and underappreciated axiomatization of Minkowski spacetime.[2] Most of us would view this as a formalization of special relativity; but since Robb preferred to *contrast* his theory with that of Einstein, let us keep the more neutral form above. In Robb's formalism one takes as primitive a class of events and one primitive relation on them: a is after b. Intuitively a is after b if and only if it is possible to propagate a causal signal from b to a, and a and b are distinct events. Robb imposes twenty-one axiomatic conditions on the 'after' relation, sufficient to guarantee that the structure of causal connectibility among events is that among events in four-dimensional Minkowski spacetime.

Robb then shows that it is possible to introduce definitions, in terms of the 'after' relationship, that are sufficient to capture such well-known features of Minkowski spacetime as timelike connectivity, null connectivity, spacelike connectivity, etc. More surprisingly, he shows that one can offer definitions of 'is a timelike inertial path,' and 'is an interval congruent to a given interval,' even for spacelike intervals, such that the spacetime structures so defined will again correspond in all their geometric features to the usual such structures of Minkowski spacetime.

The temporal asymmetry of the 'after' relationship seems irrele-

2. A. Robb, *A Theory of Time and Space* (Cambridge: Cambridge University Press, 1914). A second edition published under the title *Geometry of Time and Space* appeared in 1936. An abbreviated treatment is A. Robb, *The Absolute Relations of Time and Space* (Cambridge: Cambridge University Press, 1921). An outline of Robb's work is contained in J. Winnie, "The Causal Theory of Space-Time," in J. Earman, C. Glymour, and J. Stachel, eds., *Foundations of Space-Time Theories, Minnesota Studies in the Philosophy of Science,* vol. 8 (Minnesota: University of Minnesota Press, 1977), 134–205.

vant to his main task, and indeed it can be shown that if one starts with the symmetric relationship corresponding to 'either a is after b or b is after a,' then, following out a Robb-like construction, one can construct all those features of Minkowski spacetime that are themselves independent of a choice for the "direction of time."[3]

Now Robb, being concerned only to construct a formalized theory of spacetime, has little to say about the relationship between the observable behavior of material objects like clocks, rods, and free particles and the spacetime structures so constructed. He does tell us, however, that it is an empirical fact that light rays travel null straight lines, as he constructs them, and I assume he would take it that in our full physics we would simply add more axioms to tell us that atomic clocks measure timelike intervals relative to a reference frame, and rigid rods spacelike intervals relative to a reference frame; and that free particles travel timelike straight lines. These are totally nondefinitional on his view. We shall discuss this in more detail shortly.

A Philosophical Thesis Drawn from Robb's Formalization

Let us look at one aspect of an answer to our second question above: What important philosophical consequences can be drawn from opting for a given formalization of a theory? John Winnie has argued as follows: Grünbaum and others have asserted that the metric structure of spacetime is conventional since, even given the topology of spacetime, no particular set of metric relations between the events is singled out as preferred. Therefore the choice of metric is, in some sense, arbitrary. Hence the metric of spacetime is conventional.[4]

While Robb did not discuss the problem of formalizing the topology of spacetime, it is clear that for Minkowski spacetime, a formalization is available that takes only the 'after' relationship between events as primitive, and in which we can fully define all the topological features of the spacetime. This is certainly true in the following sense: given the 'after' relationship, and the Robbian axioms governing it, we can define what it means for an event to be in the interior of the forward (respectively backward) light cone of the event. Then we can take as the topology of the spacetime the

3. R. Latzer, "Nondirected Light Signals and the Structure of Time," *Synthese* 24 (1972): 236–280, sections I and II.
4. Winnie, "Causal Theory of Space-Time."

Alexandroff topology, i.e., the coarsest topology in which all such interiors of light cones are open. This will agree in all its structure with the usual manifold topology on Minkowski spacetime.[5]

But, Winnie continues, Robb has shown that the metric congruence structure of the spacetime is definable by the 'after' relationship alone. The same basic relation defines simultaneously the topological and metric features of the spacetime. Thus we cannot really first fix the topology and still have free choice as to the metric; hence the metric is not really conventional. If we opt for Robb's formalization we see that, at least in the context of Minkowski spacetime, the metric of the spacetime is not really a matter of "convention."

A full critique of this argument will reveal much that is of general philosophical importance. But rather than attack this thesis directly, I shall take a more circuitous route. I hope the reader will bear with me through a lengthy digression.

General Relativistic Considerations

In the first edition of *Raum, Zeit, Materie* Weyl suggested that we could, assuming spacetime to be a four-dimensional pseudo-Riemannian manifold, completely map out the metric of the spacetime in which we live by using light rays alone. The fact that the light-cone structure of the spacetime is the structure of causal connectibility suggests that 'after' is a sufficient primitive on which to found the definitions of all the metric concepts we use to characterize the spacetime. But, as Lorentz pointed out to Weyl, and as Weyl noted in later editions of his book, this thesis is wrong. For any two nonisometric spacetimes that are *conformally* equivalent (i.e., such that there is a one-to-one angle-preserving transformation from one to the other) will have the same light-cone structure.

Weyl then pointed out that if we added to our body of observational data, so far consisting of the paths of light rays, the paths of free particles, and if we assumed these free particles traveled time-like geodesics in the spacetime, then we could indeed fully determine the metric. For if there is a one-to-one mapping from one spacetime to another that preserves the paths of material particles

5. See, for example, S. Hawking and G. Ellis, *The Large Scale Structure of Spacetime* (Cambridge: Cambridge University Press, 1973), 196–197.

(i.e., the timelike geodesics), then the spacetimes are isometric, at least up to a constant factor.[6,7]

Robb's results on the definability of the metric by causal connectibility in Minkowski spacetime, and those that tell us that the metric is most certainly not uniquely fixed by the causal structure in Riemannian spacetimes, are, of course, completely compatible with each other. Suppose we know that spacetime is Minkowskian. Robb has shown us that we can determine the interval separation of any pair of points (relative to a given separation taken as unit) by means of light rays alone, light rays being taken to demarcate the boundary of causally connectible sets of events. But if we know only that the spacetime is Riemannian, then the results noted above tell us that we could not even determine that the space was flat (Minkowskian) using light rays alone, much less measure relative interval separations fully between all its pairs of points.

To clarify this point further, we might reflect for a moment on an important result of Zeeman. He shows that if spacetime is Minkowskian, any automorphism of the spacetime onto itself that preserves, in both directions, all the causal connectibility relations and the time-ordering of events that are causally connectible (i.e., if b is after a in Robb's sense, and if f is the mapping, $f(b)$ is after $f(a)$;

---

6. H. Weyl, *Space, Time and Matter*, 4th ed. (New York: Dover, 1950). See pp. 228–229 and Appendix I, pp. 313–314. See also Hawking and Ellis, *Large Scale Structure*, pp. 56–65.

7. The "up to a constant factor" qualification here is worth a moment's notice. The adequacy of using light rays and free particles alone is usually justified by arguing that to fix the constant factor is, after all, just to choose a scale of measurement. In one sense this is surely right. Even using rods and clocks to determine a metric, we need, to get actual interval values, an assignment of a specific length to a particular spatial or temporal interval—and not just congruence by transported rods or clocks. So, given our light rays and free particle paths, the full metric, including the actual intervals between events, is fully specified if we pick two events and designate their interval separation as unit. Sometimes, however, people go beyond this to claim that the notion of two distinct possible worlds which differ only in their relative scale is an absurdity. I do not believe that this follows from anything said above in this note. This is the question, not whether it makes sense to talk about the universe "doubling in size overnight," but whether it makes sense to talk about a universe in which all intervals have always been twice what they actually are. I shall not pursue this question here, but only remark that the answer is, I think, nontrivial and depends greatly upon just what possible worlds are in one's metaphysics and semantics.

and if d is after c, $f^{-1}(d)$ is after $f^{-1}(c)$) is a member of the group G, where G is the group generated by the orthochronous Lorentz group, translations, and dilations of the spacetime. In other words, if the spacetime is Minkowskian, "causality implies the Lorentz group."[8]

But Zeeman most certainly does *not* show us that any mapping between two pseudo-Riemannian spacetimes which is a one-to-one mapping and which preserves 'after' in both directions is such an extended Lorentz transformation. I think this is parallel to the fact that, while Robb shows us that in Minkowski spacetime, the 'after' relation defined inertial lines and both spatial and temporal congruence, he has certainly not shown us that we can tell which pseudo-Riemannian world we are in by using the data on 'after' relations alone. For, once again, even if Robb's axioms for 'after' are satisfied, the spacetime need not even be Minkowskian.

Now if spacetime is Riemannian, one of two possibilities holds: either the causal connectibility (or 'after') structure obeys Robb's axioms or it does not. In general it does not. In fact, unless the spacetime is conformally flat, the causal structure of the spacetime will not even be *locally* like that of Minkowski spacetime at each point.[9] And even if the spacetime is conformally flat, i.e., locally conformal to Minkowski spacetime, Robb's axioms still need not hold, for they are of a global nature; they will hold only if the spacetime is conformal to Minkowski spacetime in a global way. It is, in fact, the global nature of Robb's axioms that allows him to define the metric structure in terms of the causal when his axioms are satisfied. For, as Weyl showed, the *local* causal structure is insufficient to fix the metric even in the Minkowskian case.[10]

Now then, Robb's axioms may not hold. We allow this as a possibility, believing as we do in general relativity and therefore in the possibility that spacetime is not globally conformal to Min-

8. E. Zeeman, "Causality Implies the Lorentz Group," *Journal of Mathematical Physics* 5 (1964): 490–493.

9. E. Kronheimer and R. Penrose, "On the Structure of Causal Spaces," *Proceedings of the Cambridge Philosophical Society* 63 (1967): 481–501. See p. 483.

10. See J. Ehlers, F. Pirani, and A. Schild, "The Geometry of Free Fall and Light Propagation," in L. O'Raifeartaigh, ed., *General Relativity* (Oxford: Oxford University Press, 1972), 63–84, esp. 68–69. The original result is in H. Weyl, *Mathematische Analyse des Raumproblems* (Berlin: Springer, 1923), esp. lecture 3.

kowski spacetime. Indeed, when we look at the almost inevitably singular nature of cosmological solutions to the general relativistic field equations, it seems most probable that Robb's axioms do not hold. And if Robb's axioms fall, so does his program of defining the congruence of the metric in terms of causal connectibility.

But suppose we did believe that Robb's axioms hold. Should we accept his definitions of the metric congruence in terms of the causal connectibility relation? Two possibilities arise: (1) While we believe that Robb's axioms hold, we believe special relativity in general does not. From the general relativistic point of view this corresponds to a theory of a spacetime which, while globally conformal to Minkowski spacetime, is not flat. Empirically this is revealed to us by the fact that free particles do not travel timelike straight lines as causally defined in Robb; rigid rods do not measure Robbian spatial intervals; and atomic clocks do not measure Robbian timelike intervals. (2) Not only do we believe that Robb's axioms hold, but we believe that the special relativistic predictions about free particles, rods, and clocks hold as well.

IV SOME ALTERNATIVE FORMALIZATIONS OF ROBBIAN BUT NON-MINKOWSKIAN SPACETIME

The Maximal Robbian Formalization

Robb's axioms hold; but from the general relativistic point of view, the spacetime is believed to be non-Minkowskian. What alternative formalizations could we choose?

Option A: The maximal Robbian choice.

Here we keep as close to the Robbian analysis as possible. His postulates for 'after' remain the same. And we hold to the same definitions. Of course we must now drop such assumptions as that free particles travel timelike straight lines, that rigid rods measure intervals, etc.

Now we have some warrant for taking this approach in the history of the theory after all. In Newtonian spacetime, for example, we never assumed that particles acting only under the influence of gravitation followed spacetime geodesics. Instead we took these geodesics to be straight lines and said that the trajectories of the particles were distorted from the geodesics by the gravitational *force*. Now, of course, given the assumption that rigid rods measure general relativistic spacelike intervals and that clocks measure general relativistic timelike intervals, we must allow these "forces" (or, better, *potentials*) to influence the behavior of material objects in the spacetime in other ways as well. The gravitational force must become also a gravitational "stretching-shrinking" field.

In this light it is interesting to see how Robb himself has responded to the introduction of general relativity. He worries about how his spacetime theory is going to fare in the light of general relativity in an appendix to his *The Absolute Relations of Time and Space.* His remarks there are brief and enigmatic, but I think we can see that something like what we are proposing here as Option A, but with differences, is going on.

Robb says that although there might be some reason for speaking of particles as traveling geodesics in one of Einstein's "complicated geometries," as a *façon de parle,* "this does not imply any 'curvature of space.'" His meaning is clear: spacetime is still flat, and timelike geodesics are still straight lines. While free particles can be viewed as traveling "geodesics" in a *fictitious* Riemannian spacetime, they do not really travel as geodesics at all.

Robb also realized that in general relativity even light rays may fail to travel straight-line geodesics or even to satisfy his axioms. Here his reply is hard to comprehend. He seems to argue that there may be other causal influences that will obey his axioms even if light paths do not. In particular he thinks the propagation of gravitational influence might take the place of light in marking out the boundaries of sets of causally connectible events. Then, he says, we could still adopt his spacetime theory, taking it as an empirical fact that gravitation selects the extremely causally connectible events, and allowing light, like free particles, to be influenced by the matter-field so as to fail to follow the "real" geodesics of the spacetime.[11]

But of course this will not do. If general relativity is correct, there can even be spacetimes in which extremely causally connectible events simply do not meet Robb's postulates, inasmuch as gravitational force in a vacuum propagates with the velocity of light whether under the influence of "real matter" or in an empty non-Minkowskian spacetime. Robb should have faced the fact that in the light of general relativity, even his axioms for causal connecti-

---

11. Robb, *Absolute Relations,* Appendix, 78–80.

bility might break down, and with it his whole program for a theory of spacetime.

But our modest Robbianism is still worth examining. Once again, it states: if at least the Robbian axioms for causal connectibility hold up—i.e., if the spacetime is at least globally conformal to Minkowski spacetime—we can retain a formalization of our spacetime theory that preserves Robb's axioms and definitions, even for such things as timelike geodesics (inertial lines) and spatial and temporal congruences, and make our theory fit the facts by changing the postulates that connect the behavior of free particles, rigid rods, and clocks with the spacetime structure. That is, we can introduce gravity as a "potential" superimposed on the spacetime.

A Formalization Differing in Definitional Consequence Only

Option B: Choose a formalization that agrees with Robb's on the status of terms and on those consequences framable in only the (mutually agreed) primitive terms, but that differs from Robb on the definitions offered, and on definitional consequence generally. What would this look like? First of all the formalization would take as primitive only those terms designated as primitive by Robb. And any proposition involving only the primitive terms would be a consequence of the new formalization if and only if it were a consequence of Robb's formalization. But the formalization would differ from Robb's, for example, by defining 'timelike geodesic' as 'path of a free material particle.' And whereas in Robb's formalization free particles would not travel timelike geodesics, rigid rods would not measure spacelike intervals, and clocks would not measure timelike intervals, in this formalization the usual relativistic assumptions about the relationship between these material objects and these spacetime structures might hold.

But the Robbians and the proponents of this new spacetime theory *might* very well agree that their two formulations were formalizations of the same theory, although I do not know if Robb himself would have accepted this line. How would they argue? Perhaps as follows: "We agree about the primitive consequences of our formalizations. And the primitive consequences of a formulation contain the totality of its *factual import*. We *choose* to define some nonprimitive terms in different ways, but that is a free choice in any case, subject only to the formal conditions of proper definability. Having chosen these differing definitions, it is hardly surprising that we extract from our axioms *apparently* incompatible consequences in cases in which they are propositions containing defined terms. But this apparent incompatibility is only apparent. To think that there is real incompatibility is to fall prey to fallacies of equivocation induced by failure to pay proper attention to the differing *definitions* we offer of the nonprimitive terms. Our formalizations, despite their differing linguistic appearance, really are just different ways of expressing one and the same theory."

We noted above that what we called the maximal Robbian formalization had the "advantage" that in the case in which Robb's axioms were satisfied but in which not all of special relativity was correct, this formalization made gravity play a role similar in some ways to its role as a force in Newtonian mechanics and in special relativity. The advocate of this new formalization might argue like this about the earlier physics: "It is true that in pre-general-relativistic physics we treated gravitation as a force superimposed on the spacetime. In my formalization, factually and observationally equivalent to Robb's though it is, we *talk* as if gravity really were curvature of spacetime. This is a more natural way of talking than Robb's, for my way maintains the connection between rigid rods and spatial separations, and between clocks and temporal separations, which existed in the prerelativistic theory and which is preserved both in the usual formalizations of special relativity and in the usual formalizations of general relativity.

"Of course in my formalization we speak of particles acted on only by gravity as following curved timelike geodesics, and we speak of curved spacetime, even though we did not talk this way in the pre-general-relativistic theory. But we must realize that that older way of talking was *only* a way of talking. For just as we have a choice between the Robbian-Minkowski spacetime plus gravitational potential fields, and the curved spacetime formalization, to formalize our present theory of spacetime, which — in general relativistic terms — is globally conformal to Minkowski spacetime but not flat, we had the same options in pre-general-relativistic physics. For example, as Trautman has shown, we can formalize Newtonian physics from the standpoint of curved spacetime just as the maximal Robbian shows us we can formalize our present theory from the standpoint of flat spacetime. In the curved spacetime formalization of Newtonian physics, the same natural association of rigid rods and clocks with the appropriate intervals is maintained. But we now talk of free particles — those acted upon only by gravity—as 'following the timelike geodesics of a curved space-time.'[12]

"But we must note that in both these cases it is only a matter of *how we talk,* not *what we believe.* For in both cases the observational (factual) consequences of the alternative formalizations are the same. This follows from the fact that the formalizations agree on the primitive and defined status of terms and propositions and on primitive consequences, differing only on definitional consequence."

Such a thinker might continue to defend a preference for his way of formalization as follows: "While both the maximal Robbian and my formalization of this spacetime theory are based on one and the same theory, mine is superior to his in a number of ways, even though it differs from his merely in the definitions adopted for the (agreed) defined terms. The only advantage of the Robbian's formulation is that in his theory, as in the usual Newtonian theory, particles acted upon by gravity do not travel timelike geodesics, but instead are acted upon by forces. As I have just pointed out, this "continuity" with pre-general-relativistic physics is not a critical point, for we could, if we wished, even reformulate Newtonian physics to meet the general relativistic way of talking, in which particles acted upon only by gravity follow timelike geodesics in a curved spacetime.

"My way of speaking has other continuities, both with pre-general-relativistic physics and with general relativity, which the maximal Robbian's formulation lacks. While Newtonian physics, special relativity, general relativity, and my formulation all speak of rods and clocks as correctly measuring spacetime intervals, the maximal Robbian drops this common way of speaking for the sole reason that this is the only way he can keep his Robbian definitions of congruence, etc.

"My view is further shown to be preferable by the following reason: I believe that Robb's axioms for causal connectibility hold. For they hold both in the maximal Robbian formalization and in mine. But I know about general relativity, so I realize my present theory may be wrong, so wrong that the Robbian axioms really do

12. A. Trautman, "Comparisons of Newtonian and Relativistic Theories of Space-Time," in B. Hoffman, ed., *Perspectives in Geometry and Relativity* (Bloomington: Indiana University Press, 1966), 413–425. See also C. Misner, K. Thorne, and J. Wheeler, *Gravitation* (San Francisco: Freeman, 1973), chap. 12, "Newtonian Gravity in the Language of Curved Spacetime."

not hold. Should I discover this, I shall have to change my theory. Now the maximal Robbian will not only have to give up belief in the truth of his causal connectibility axioms, he will have to drop all his definitions of congruence, timelike geodesic, etc., as well, for the very formulation of these definitions depends upon the Robbian axioms holding.

"I, on the other hand, shall, of course, have to drop my Robbian axioms of causal connectibility. But I shall be able to preserve my definitions of timelike geodesics and congruence intact, for they do not depend, in my formalization, on the Robbian axioms, and they correspond exactly to the usual general relativistic definitions of these spacetime terms, definitions which are adequate even when the spacetime is not globally conformal to Minkowski spacetime. My formalization, then, while admittedly differing only definition-ally from the maximal Robbian's, is more continuous in its way of talking—both with pre-general-relativistic physics and with general relativity in their usual formulations—than is the maximal Robbian's way of speaking. Thus one should opt for my formulation." Here the reader will undoubtedly recall the familiar thesis of Eddington, Schlick, and Reichenbach: as long as two formalizations have all the same "observational consequences," they are the formalizations of one and the same theory, differing only in linguistic formulation.[13]

At this point several crucial issues about definition in formalizations have finally surfaced, and we must now make an initial assault upon them. The reader should be assured that our tentative probes here will hardly be the last moves we shall make on these issues, for we are now about to discuss fundamental problems which will appear again and again as we examine particular formalizations of particular theories, their similarities and differences, the reasons for and against adopting them, and the consequences of doing so.

In the rationalization given above for claiming that the two formalizations were based on the same theory, a fundamental proposition was assumed: identity of primitive consequences constitutes identity of observational and of factual consequences. But why should anyone assume this? I think the answer is clear: in both formalizations the choice of primitive or of defined status for a

13. See H. Reichenbach, *The Philosophy of Space and Time* (New York: Dover, 1958), chap. 1. See also L. Sklar, *Space, Time, and Spacetime* (Berkeley: University of California Press, 1974), chap. 2.

term, and the resulting classification of consequences of the formalizations into primitive and definitional, were *designed* to relegate the totality of observational and factual consequences of the formalizations—these two classes being *assumed* to be coextensive—to the class of primitive consequences.

First of all, the argument assumes that a consequence of a theory cannot be factual unless it is, at least in principle, observably either the case or not the case. We accept that a theory cannot entail any putative "facts" unless they are putative observable facts; for, in the best verificationist vein, we accept that a "factual" difference with no observable consequence is only an *apparent* factual difference. Second, the argument assumes that the classification of terms and propositions into primitive and defined is not an arbitrary classification, but one designed to mark out in the margins of the formalization those parts of its contents that are accessible to empirical *test* and those that are open to "arbitrary" linguistic *choice.*

What should count as a primitive term in a formalization? The frequent answer is that primitive terms should be only those whose applicability or nonapplicability to a situation can be determined without in any way presupposing the truth of the theory in question. After all, the fundamental axioms of the formalization are to be presented as framed in the primitive terms only. And it is the empirical test of the correctness of these axioms that will tell us whether or not the theory so formalized is correct. But if we could not determine the applicability of a primitive term to a situation without presupposing the correctness of the theory in question, how could empirical testing ever get under way?

One version of this thesis would have us restrict the primitive terms to those whose applicability or nonapplicability is in some sense a matter of "immediate sensory awareness." Primitive terms are then the basic observation terms of some radically empiricistic philosophies of science. This is certainly Robb's position with respect to his 'after' primitive. For, he says, the grounds for choosing 'after' as a primitive are that we can be *directly conscious* of some pair of events—that is, a single "consciousness" can—and this same single consciousness can immediately, without any theoretical presupposition or inference, tell that one of these events in the pair is after the other.

The reader acquainted with some history of philosophy who reads Robb's introductory sections on the status of 'after' and on the "locality of simultaneity," and who encounters Robb's belief that simultaneity for events spatially separated is nonsensical, will recognize a close similarity between Robb's position and that expressed in the third chapter of Bergson's reply to special relativity, *Duration and Simultaneity*. Bergson's work has been, I think, unjustly neglected. This is largely his own fault, for many readers have been dissatisfied with his account of relativity because of the irrationalism rampant throughout Bergson's work, and because of the latter parts of *Duration and Simultaneity,* which constitute a confused attempt to refute the so-called clock paradox in special relativity.[14]

This rationale for choice of a primitive leads immediately to philosophical trouble: it looks as though our physics is going to be founded on a solipsistic, private, subjective observation basis. Reichenbach, in his well-known axiomatization of special relativity, was well aware of this problem, even though elsewhere in his work he talks more like Robb and seems to take the primitive as the "directly apprehendable by a consciousness." In his book on special relativity Reichenbach uses, for example, *coincidence* of events as a primitive. But, he says, by this I mean *physical* coincidence, not coincidence of "appearances" in some subjective consciousness. Reichenbach calls those propositions that are framable entirely in terms of his primitives, *elementary facts.* What makes them elementary is not immediate apprehendability by consciousness, but rather the fact that they "remain invariant with respect to a great variety of interpretations." For example, we perform the Michelson-Morley experiment. We might have many different physical interpretations of the results. But all such theoretical accounts will agree as to whether the physical interference lines did or did not remain coincident with designated marks when the interferometer was rotated.[15]

As it stands, this definition of elementarity leaves something to be desired. For surely I could have considered a theory in which I explained away the *apparent* null results of the Michelson-Morely experiment, not as was done, by assuming the null results were a physical reality and by choosing one of the alternative accounts for them (Lorentzian aether theories or special relativity, for example), but by explaining the results away as a subjective illusion somehow

14. Robb, *Geometry,* Introduction. H. Bergson, *Duration and Simultaneity* (Indianapolis: Library of Liberal Arts, 1965), chap. 3.
15. H. Reichenbach, *Axiomatization of the Theory of Relativity* (Berkeley: University of California Press, 1969), Introduction.

to be accounted for on the basis of a theory which no one has ever proposed and which I cannot even now begin to construct.

If primitiveness is to be identified with elementarity in Reichenbach's sense, then, rather than with direct apprehendability in Robb's, I think Reichenbach's definition in terms of "relative invariance with respect to a great variety of interpretations" must be supplemented to read "relative invariance with respect to the great variety of interpretations *which one has in mind as plausible theoretical accounts of the experimental observations.*" The reader may now begin to see the point of the title of this chapter. I am suggesting that even to begin to characterize the basis on which we shall choose either to accept a theory or formalization or to reject it in favor of some alternative, we need some preliminary *assumptions* about the range of theories we are going to consider.

Many philosophical readers will at this point recall the perennial doctrine of the "theory-ladenness of all terms including the observational" which recurrently makes its mark on the philosophy of science. Indeed, issues of this kind are what I have in mind. But I shall reserve my more general philosophical arguments about this issue and try to outline those aspects of the general philosophy behind the primitive/defined distinction in formalizations, which we need in order to continue our examination of the concrete dispute we are enmeshed in.

If we are presented, then, with a formalization in which some terms are taken as primitive, we should probably assume these are the terms of the theory whose applicability or nonapplicability to a given physical situation is independent of the acceptance of his theory. Naturally the defined terms are supposed to have just the opposite status. If the axioms of the theory are correct, then certain physical situations exist. We can characterize these in terms of the primitive alone. But it may be more convenient to introduce new single terms to refer to some of these physical entities or relations, so we introduce definitions and defined terms.

Just as restricting the axioms to propositions framable only in the primitive vocabulary was supposed to have the advantage that we could *test* the axioms for correctness or incorrectness without presupposing the theory correct, the definitions are supposed to have the opposite virtue of not being amenable at all to physical test. Since the defined terms are *introduced* by the definition, and *stipulated* to hold only when the defining situation characterized in the primitive vocabulary holds, there can be no question of *testing* the correctness of a definition, although it is, of course, a testable question whether the definition is "admissible," since the existence and uniqueness propositions necessary to legitimatize it as a definition are themselves expressible in the primitive terms alone. And since the primitive vocabulary exhausts those terms whose applicability or nonapplicability is decidable independent of accepting the theory, the totality of empirical tests of the theory resides in drawing from it its primitive consequences and testing them.

We are now in deep philosophical waters indeed, and this will hardly be the last we have to say about these matters. But we have enough, I think, to understand at last why our maximal Robbian and his opponent can agree about a number of issues. They agree as to which terms are primitive and which defined in their formalizations, i.e., their formalizations are alike in the status they attribute to terms. They agree that their formalizations are based on the same theory. They disagree only with regard to definitional consequences of their theories; but they agree that these differences are simply the result of having *chosen* differing definitions for some terms, and that these differences are in no way a mark of *empirical* disagreement.

A Formalization Differing Only in the Attribution of Definitional Status to Consequences

Let us now return to our concrete case. We are considering the options available to us in the situation in which, from the general relativistic point of view, we believe that spacetime is globally conformal to Minkowski spacetime but not flat. Physically this means we believe that whereas Robb's axioms for causal connectibility hold, special relativity does not, because it fails on the "matter" side. Free particles, rigid rods, and clocks do not, we believe, correspond in their behavior to the geometry constructed using light (or causal connectibility or 'after') in Robb's manner as they would if special relativity were correct.

Here is another option we could take in formalizing our beliefs:

Option C: We adopt a formalization that agrees with the maximal Robbian about the primitive or defined status of the terms. It also has exactly the same consequences, primitive and definitional, as the maximal Robbian formalization. But it allocates definitional and postulational status to some of these consequences in a different way.

In this formalization one would agree, for example, with the maximal Robbian that free particles do not travel inertial lines, that rigid rods do not measure spatial intervals, and that clocks do not measure timelike intervals. But the formalization might declare that what *defines* an inertial line is still fixed by that consequence that relates the behavior of free particles to inertial lines. And the formalization would then declare that what the maximal Robbian takes to be a causal *definition* of inertial line is a *postulate* about the relation of inertial line structure to causal structure.

As one might imagine, the proponent of this formalization will usually say he is most certainly formalizing the same theory that the maximal Robbian is formalizing. He is even using the same words, and he believes the same terms to have their criteria of applicability independent of accepting the theory. But, he says, since his definitions are the maximal Robbian's postulates, and vice versa, he is disagreeing with the maximal Robbian about the *meanings* of at least some of the defined terms.

Why would anyone prefer such a formalization to the maximal Robbian's? In this case, I doubt that anyone would, but later we shall see a case in which just such a preference is plausible. The proponent of this formalization might argue that his is preferable to the maximal Robbian's simply because the meanings he gives to the defined terms are closer to "what they ordinarily mean" than are the meanings the maximal Robbian gives to them.

He might argue like this: "Our ordinary meaning of 'inertial line' is that it is something fixed by the motion of free particles. And our ordinary meanings of spatial and temporal 'intervals' is that they are given by rods and clocks. The maximal Robbian instead defines these quantities in a causal way. Now I do not say that he *cannot* so define these terms. Nor do I claim that there is anything physically wrong with the theory he formalizes in the peculiar way that results when one gives such novel meanings to the terms. I am only saying that my definitions, my meanings, are closer to what the terms 'meant all along.' "

The argument is not very persuasive here. For while it may be "giving a novel meaning" to 'inertial line' to define it causally as the maximal Robbian does, the meaning given this expression by the proponent of formalization of Option C is peculiar as well. For according to him, in agreement with the maximal Robbian and in opposition to Option B discussed above and Option D discussed below, free particles do not travel inertial lines. Now if 'inertial line' had an ordinary meaning it was, I suppose, 'path traveled by a free particle.' So the proponent of Option C is offering a definition of 'inertial line' just as peculiar in its own way as the maximal Robbians. As I noted, we shall later see a case in which just such a move from the Robbian causal formalization to an alternative seems more plausible.

Another possibility is this: someone might agree with the maximal Robbian that only causal connectibility should be taken as primitive. He might, however, be reluctant to assign "definitional" or "postulational" status among the sentences of his formalization which involved terms other than the primitive terms. In a sense he would be unwilling to assign propositions in formalizing the role uniquely, either of fixing meanings or of stating facts. Those who eschew the analytic/synthetic distinction would argue in this way, I believe.

Nonetheless, such a thinker might still agree with the maximal Robbian that both formalizations have all the same "factual" consequences—assuming, that is, that they have the same primitive consequences. The idea here would be of a formalization that, like that of Option B, was "merely another way of formalizing the same theory the maximal Robbian formalized," superior in that by refusing to allocate definitional vs. postulational status to the propositions of the theory which involved vocabulary other than the primitive one, this formalization permitted us a more flexible way of talking about theory change and a more realistic way of talking about meaning.

Basically, this position would look like this: "After all, the real import of a theory is, indeed, its observational consequences. My formalization, like the maximal Robbian's, has all the right observational consequences, and these are captured in the primitive consequences that my formalization shares with the maximal Robbian's. He, however, has taken the further—and gratuitous and misleading—step of allocating "analytic" and "synthetic" status to the propositions of his theory which involve nonprimitive terms. Since this has no empirical consequence, he shouldn't do it."

On this account, while it is important to demarcate the observational import of a theory, further discrimination among its consequences by marginal notes about their "analytic" or "synthetic" status are pointless and, from the point of view of an adequate account of meaning, misleading. But I shall not burden the reader with one more attack on the analytic/synthetic distinction, for I think we have more relevant fish to fry.

A Formalization Differing Only in One-Way Definitional Consequence

Suppose someone offered a formalization of a theory to account for the spacetime of this world in which he, the maximal Robbian, and the proponent of Option B above all agreed that this new formalization differed from the earlier two in a primitive consequence. Then all would agree there was no sense in saying that this was a new formalization of the theory formalized alternatively by the maximal Robbian and the proponent of Option B. Instead they would agree that this was the formalization of a new theory incompatible with that earlier formalized.

But suppose the following peculiar situation arose: a new formalization is proposed. Let us call it Option D.

Option D: Here we are presented with a formalization, some of whose consequences differ from those of the maximal Robbian. But this formalization also differs from the maximal Robbian's in the status it attributes to some terms. In particular, at least one consequence of this new formalization is incompatible with those that follow from the maximal Robbian's formalization, and it has the following interesting feature: whereas the maximal Robbian declares this consequence to contain a defined term—defined, at least, in *his* formalization—the new formalization, with its new status attribution to terms, declares the consequence to be *primitive.* Further, let us suppose the maximal Robbian declares no consequences of the new formalization primitive but incompatible with the consequences of his own.

Let us look at the situation first from the point of view of the maximal Robbian, and then from the point of view of the exponent of the new formalization. The maximal Robbian says: "This is just like the situation discussed in Option B above. This new formalization differs from mine only in definitional consequences, for the only consequences of it "incompatible" with consequences drawable from my formalization are those containing what *I* take to be defined terms. So this is really a new formalization of the same theory as mine, except that this formulation chooses to express the theory in a different way. If there are grounds at all for choosing among these formalizations, they are just like the grounds for choosing between my maximal Robbian formalization and Option B above. They are choices of expression." (Incidentally, again, I am not sure that Robb himself would have tolerated this move.)

But now let us see what the proponent of this new formalization has to say: "My formalization has consequences incompatible with those drawn from the maximal Robbian formalization. Furthermore, *I* say that these consequences, framed in what *I* take to be primitive terms alone, serve as grounds for testing the theory. They contain no terms whose applicability or nonapplicability requires presupposing the truth of the theory. Clearly what I am proposing is not a new formalization that merely "rephrases" the theory formalized by the maximal Robbian; I am proposing a new theory to account for the data *incompatible* with the theory formalized by the maximal Robbian."

Why would anyone declare that a term taken by the maximal Robbian to be defined, is really a primitive term? This is a subtler issue than it appears to be at first sight. One reason he might have for disagreeing with the maximal Robbian on this issue might be this: he might be taking primitive terms to be those whose applicability or nonapplicability is determinable by "immediate observation," and then saying there are more features of the world accessible to immediate, totally theoretically unmediated, test than the maximal Robbian allows. He might argue, for example, that the maximal Robbian is wrong in thinking that only facts about causal connectibility are directly accessible epistemically. He might assert, for example, that he can immediately tell whether or not two noncoincident but nearby rigid rods are congruent, just as Robb asserts that we can immediately tell when one of two events in our consciousness is "after" the other. If that is his line of reasoning, we are in for one of those endless debates about just what is accessible to immediate observation, and about whether, if we insist upon immediate apprehensibility as a criterion of primitiveness, we ever can construct a science that breaks the "veil of perception" and becomes more than a lawlike systematization of a solipsistic consciousness.

On the other hand, he might argue like this: "I don't take 'primitive' to mark out 'determinable by immediate inspection and so totally independent of theorizing.' But I do have a notion of primitiveness which is important in that it captures "factualness" in the following way: propositions framed entirely in primitive terms state *facts*. They are in no way "conventional." Of course it is in some sense convention that the words mean what they mean, but when I say that a term in a formalization is primitive, I am indicating that a consequence of this formalization framable in these terms alone is true or false irrespective of the way in which terms gain meaning only from the role they play in *this* theory."

Whichever position he takes about primitive terms, he will continue like this: "While the maximal Robbian, for example, says that free particles do not travel timelike geodesics, I say they do. While the proponent of Option B also says they do, he says he is disagreeing with the maximal Robbian only as to the meaning of 'timelike geodesic,' a term whose meaning, for both of them, is fixed by the differing definitions in their respective formalizations. But I say we should take 'timelike geodesic' as a primitive term in our formalization; I disagree with the maximal Robbian about a *fact* of the world (whether free particles do or do not *really* travel timelike geodesics). Furthermore, while the proponent of Option B *thought* he was disagreeing with the maximal Robbian only about 'meaning of terms,' he too is proposing a theory *incompatible* with the maximal Robbian's. The incompatibility, however, was disguised by his *mistaken* attribution of defined status to 'timelike geodesic.'"

As we know, some would even deny the usefulness or intelligibility of any observational/nonobservational or "factual"/"conventional" distinction altogether. I think we shall be able to consider them, without loss of content, as advocating that all the terms of a formalization of a theory should properly be taken as primitive in our sense.

Variations on Option D

The exponent of Option D is most understandable when he takes the first line of reasoning we have proposed for him: he believes in "pure observation languages"; he believes in terms whose applicability or nonapplicability to a situation is totally theory-independent. He disagrees with the maximal Robbian and with the proponent of Option B only about which terms are really in the "pure observation language."

The exponent who takes the other option clearly has his work cut out for him. Just what is this theory of meaning, and just how do words get their meaning according to it? How does he rationalize a useful distinction between primitive and defined terms in a formalization if he can't pick out the former by their "pure observational status"? Perhaps he has some kind of theory in mind which, like that implicit in the Reichenbachian notion of "elementary fact" noted above, rests upon a view of the role of antecedent theories (and theories one is entertaining but not believing) in fixing the meaning of the terms in the theory we are as a matter of fact proposing. I shall return to these questions, if not answer them. But let me note here a few of the issues that arise when one reasons in this way. While I shall not try to develop the theory of *meaning* required by this position, I shall try to say something about another problem faced by the individual who takes such a line.

This individual proposes that his formalization is based on a theory *incompatible* with that of the maximal Robbian. But he agrees that both his formalization and the maximal Robbian's give rise to all the same consequences about causal connectibility. Now he might consider that Robb is even wrong in thinking that the facts about causal connectibility were somehow "immediately accessible to observation in a totally theory-independent way." We shall, indeed, explore this issue later in some detail. But for the moment suppose he grants that at least those propositions are open to immediate empirical determination. Still, he says, I take our theories to be incompatible because they conflict on other propositions which, while I admit them to be *not* directly testable empirically without presupposing the correctness of the theory, I still take them, unlike the maximal Robbian, to be *factual* and not in any sense a matter of conventional choice about the meaning of words.

But, one immediately asks, how could one possibly decide which theory was correct? For example, how could one decide whether or not free particles follow timelike geodesics, given that this individual proposes that this is a *factual* question—not in any way a definitional one—and yet tells us we cannot decide its correctness by any "immediate empirical" means? Of course our maximal Robbian has no problems here, for he says these are really just different formalizations of one and the same theory. But our individual continues to deny this.[16]

---

16. For further treatment of the matters which follow immediately, see Sklar, *Space, Time, and Spacetime*, 119–146.

One thing he could do is opt for skepticism: "I don't really believe free particles travel timelike geodesics, unlike the maximal Robbian, who says they do not. All I wanted to do was demonstrate to you that there were other theories besides maximal Robbianism that were true to all possible observational facts (if maximal Robbianism is) and yet incompatible with maximal Robbianism. Actually I believe you haven't the faintest reason for choosing one over the other of these theses about free particles. Of course only one of these theses can be true, but you simply can never know which." Skepticism.

Another thing he could do is opt for Poincaréan conventionalism, whatever that is. He could say: "Here we have two incompatible theories with all the same possible observational consequences. No empirical test could ever induce preference for one over the other. So we must *conventionally choose* which is true. We might even go so far as to say "Truth just *is* a matter of convention." Notice, however, that I am not claiming that our choice is merely a matter of choosing "how we talk." That is what Eddington, Schlick, and Reichenbach believe. I am claiming it is a choice among *incompatible,* but empirically indiscriminable, beliefs."

Another position he could opt for is a version of apriorism: "In choosing which theory to adopt we must always use, above and beyond the empirical evidence in favor or against a theory, an initial a priori plausibility of the theory. Without such 'believabilities,' intrinsic to the theory in question and independent of the empirical evidence for or against it, theoretical decision-making could never get under way. So even if two incompatible spacetime theories have all the same observational consequences, there may still be good reason, on an a priori basis, to adopt one rather than the other."

Or he might assert that the choice of theory is always to some degree motivated by "continuity with previous theory," and that considerations of this kind could motivate his choice. He might say: "Suppose I have two incompatible spacetime theories that are observationally indistinguishable. One of these theories may still be more "continuous" with the older theory it replaces. That is, it makes less of a change in our beliefs to drop the old theory for this new alternative than for the other alternative. In this circumstance it is more reasonable to believe the new hypothesis, which is minimally different from our older rejected theory, than it is to believe the alternative more 'radical' theory."

It has been alleged, for example, that even if general relativity is really incompatible with the Reichenbachian alternative of flat spacetime and universal forces, and even if these two alternatives are observationally indistinguishable, we should still opt for general relativity since it is "more continuous with" or "a more conservative change from" the older theories of gravitation. Since this philosophical claim has been made with some frequency, it might be worthwhile to comment on it here.

There are, I think, two problems with this approach:

(1) First of all, each of the new alternatives can be, and usually will be, "more conservative" with respect to the antecedent theory in some respect or other. How can we decide which *way* of being conservative should take precedence or otherwise "weight" the value of being conservative in some particular respect? For example, general relativity is more conservative than the universal force hypothesis in that in the former case, clocks and rods at different spacetime locations are congruent if they are congruent when brought into coincidence. And this is the way things were in Newtonian and special relativistic physics.

But in Newtonian physics, and perhaps in what would have been a "natural" theory of gravitation from the special relativistic point of view, particles acted upon by gravity do *not* travel timelike geodesics. They are not "free" but "forced." This remains true in the universal force hypothesis, but fails to be true in general relativity. Crudely, the universal force hypothesis is more conservative than general relativity in that it, like Newtonian mechanics and special relativity, is a *flat* spacetime theory, whereas general relativity is not.

Which way of being conservative with respect to the older theory is more important? Why?

(2) A second difficulty with the approach is this: we are told to be as conservative as possible with respect to antecedent theory. But suppose, when we have considered the options available to us for the new theory, we go back over the older theory and realize that it too could have taken another form observationally indistinguishable from the form it had. We might end up with this view: one option for our new theory is most conservative with respect to antecedent theory as it was. But the other option is most conservative with respect to another theory we could have had earlier, which

is observationally indistinguishable from the older theory we did
have.

Suppose, for example, we decide that the universal force theory
is most conservative with respect to Newtonian mechanics because
it, like Newtonian mechanics, is a flat spacetime theory. But then
we discover, reflecting on general relativity, something like Trautman's curved spacetime version of Newtonian mechanics. And
suppose general relativity is most conservative with respect to this
now revised antecedent theory!

The individual whose "conservative" doctrine we are now examining could avoid this second problem only by becoming more
conservative. He would have to say: "I now realize that all along
there was an alternative theory that was observationally indistinguishable from my older theory. But I hold to the following conservative maxim: if you believe a hypothesis, there is no good
reason to drop your belief just because you discover that there is
another hypothesis which is just as good as, but no better than, the
one you hold. I now use this retrospectively to justify my taking the
older theory, which I actually had as preferable in believability to
these newly discovered alternatives to it, and as being the correct
starting point for conservative modifications of theory."[17]

But another problem arises for this position in the present context. We may say that if a Trautman had come along in the eigh-
teenth century, a good conservative would have been justified in
not dropping his Newtonianism and replacing it with this alternative theory. But now, given that we accept general relativity, might
it not be rational for us, even if we are generally conservative, to
assert that the theory that should have been held in the nineteenth
century was Trautman's curved spacetime Newtonian mechanics,
and not Newton's flat spacetime theory? It is one thing to be conservative with respect to physics as it was, but quite another to be
conservative with respect to physics which, we *now* see, would have
been as it should have been.

Let us consider a last option for our individual who tells us that
theories may be incompatible even if observationally indistinguishable, and who wants to tell us which of such alternative
theories we should adopt. Suppose he argues like this: "I believe
that Robb's axioms for causal connectibility hold. But I am well

17. See L. Sklar, "Methodological Conservatism," chap. 1 in this volume.

aware that I might be wrong about this. Now I believe that free particles do not travel Robbian timelike geodesics, for I believe that the matter axioms of special relativity do not hold. Suppose I found out that, contrary to my present view, Robb's axioms failed. What would I do? The only theory I have available to apply in that case is general relativity. In general relativity Robb's axioms frequently — in fact, usually — fail to hold of the world. But free particles do still travel timelike geodesics in this curved spacetime theory. Therefore, looking toward the possible worlds I now envision, and considering the possible theories I have in my repertoire to describe these worlds, I argue as follows: here are two incompatible theories that are observationally indistinguishable. One is the maximal Robbian, in which free particles do not travel timelike geodesics; the other is the version of general relativity suitable to a world which is globally conformal to Minkowski spacetime and in which Robb's axioms hold but free particles do travel timelike geodesics. Which should I believe?

"I say I should believe the latter. Why? Because if I discover that, contrary to my present beliefs, Robb's axioms do not hold, I could, with this latter choice move easily and conservatively to a new theory — a general relativistic account of a world not globally conformal to Minkowski spacetime. In this theoretical transition I could hold much invariant. I could still believe, for example, that free particles travel timelike geodesics. If I adopted the maximal Robbian theory, however, and then discovered that Robb's axioms do not hold, I would need to change my theoretical beliefs more radically.

"I am assuming the world is one of the possible general relativistic worlds. Which one is it? I believe it is one in which Robb's axioms hold. But I might be wrong about this. So, given that I believe that they do hold, I should believe that theory which is one of a spectrum of possible theories I have in mind, one of which will hold even if some of my present assumptions fail.

"I wish to believe a theory in which Robb's axioms hold. But which of the many incompatible theories in which they hold should I believe, given that I think there are many that are observationally indistinguishable? I should believe that theory which (1) is a specialization of a general class of theories, one of whose members I would adopt if I no longer believed that Robb's axioms do hold; and (2) is selected because it is the one member of this class in which

Robb's axioms do hold and which is compatible with the observational data.

"Now the class of theories I have in mind is the class of general relativistic models of the world. The theory in this class in which Robb's axioms do hold is one in which free particles travel timelike geodesics, rigid rods measure spatial intervals, and clocks measure timelike intervals; i.e., given our data, the theory is the general relativistic model of a world globally conformal to Minkowski spacetime. Now I believe the world is not flat. So I believe the free particles will not travel the timelike geodesics of the maximal Robbian theory. Therefore I should choose the theory that says free particles do travel timelike geodesics, and in which the maximal Robbian's timelike geodesics are not really timelike geodesics at all; I should not choose the maximal Robbian's theory, in which his timelike geodesics are really timelike geodesics but free particles just don't travel them.

"I make this choice *not* on the basis of my observation, for the maximal Robbian's theory and mine predict the same observational consequences. Rather, I make this choice because my theory is a special instance of a class of similar theories, one of which I would adopt even if I found out, contrary to my present belief, that one of Robb's hypotheses about causal connectibility is wrong; and the maximal Robbian's theory does not naturally fit into a class of theories at least one of which is still available to me even if I were to find out that some of my present assumptions about the world are wrong."

This individual is saying: theories can be genuinely incompatible even if observationally indistinguishable. But I may still have grounds for preferring one over the other. The grounds are not a priori plausibility, nor even conformity with antecedent theory. They are rather the natural or unnatural role which the theory plays in the light of my assumptions about the other possible theories of the world which *might* hold — although I do not believe they do — and which I *would* adopt if I found out that some of my assumptions about the world are false.

Once again, it is asserted that theoretical decision-making depends upon some *assumptions* we make about the spectrum of possible theories we are likely to consider, and not just upon the "observational facts." We shall return to this point in some detail when we consider formalizations of general relativity later in this chapter.

A Formalization Differing Only in the Status Attributed to Terms

I noted earlier four ways in which formalizations can differ from one another, but I have discussed only three alternatives to the maximal Robbian formalization. I have saved the fourth, Option E, till last, primarily because it is the one option which, to my knowledge, no one has yet espoused. Option B would be espoused by Eddington, Schlick, and Reichenbach, I believe, on the grounds that it is "descriptively simpler" than the maximal Robbian formalization. Option C would be espoused by someone who thinks the consequences of maximal Robbianism perfectly all right as they stand, but who has a different predilection for attributing definitional or postulational status to them. A Quinean, for example, who wishes to eschew such marginal status attribution altogether would adopt a version of Option C. Some version of Option D would probably be espoused by someone who, looking at the data in the light of general relativistic possibilities, and disagreeing with Robb's insistence that only causal connectibility should count as a primitive notion, would interpret the data to mean that the spacetime of the world, although globally conformal to Minkowski spacetime, was *really* not flat at all.

Here is Option E: Choose a formalization exactly like the maximal Robbian, except that different terms are designated 'primitive' and different terms are designated 'definitional.' This could be done in two ways: either (1) one might take causal connectibility as definitional instead of primitive; or (2) one might take some additional terms over and above 'causally connectible' as primitive. Since the latter is the more likely move, let me discuss what might motivate it.

Someone making this move might argue: "Robb is wrong to take causal connectibility as the only observational (or, perhaps, factual) feature of the world captured by a theory of spacetime. For example, congruence for spatial intervals at a distance is also observational (factual), as is being an inertial line. But the maximal Robbian is correct in thinking that the spacetime of the world is flat, that free particles do not travel inertial lines, etc. What the maximal Robbian says about the spacetime of the world and its connection with material objects is correct; he is only wrong in thinking that some of these consequences of his formalization are matters of definition when they are really matters of observation (or fact)."

It is no surprise that no one has ever offered an approach like this. Who, faced with empirical facts that would normally lead a general relativist to say that spacetime was nonflat but globally conformal to Minkowski spacetime, would ever assert that the spacetime was flat, and not as a matter of "definition" or "convention" — i.e., not because we could, if we wished, reformulate the general relativistic theory as a flat spacetime theory, but because we could observe that this was really the case? Or, even if we couldn't decide this observationally, because there were other reasons for thinking this Robbian account both truly incompatible with the general relativistic account and factually correct?

I shall forgo consideration of the other approach, with its denial of primitive status to causal connectibility, until later, when we shall examine that approach in another context.

Summary of Part IV

Let me summarize what has been going on in the last few sections. I have been exploring the desirability or undesirability of the Robbian definitions of such notions as inertial line (timelike geodesics) and spatial and temporal congruence, by subjecting Robb's theory to stress. Maximal stress consists in saying: suppose Robb's axioms for causal connectibility do not even hold. Under that stress his method of definition clearly breaks down. But then I have said: let us subject Robb's theory to less, but still to some, stress. Let us assume we believe that his axioms of causal connectibility hold, but that the postulates that would supplement his theory if special relativity were correct break down. That is, let us assume we believe that free particles do not travel Robbian inertial lines, rigid rods do not measure Robbian spatial congruence, and clocks do not measure Robbian temporal congruence.

Applying this stress to Robb's theory, we have seen that the following options can be delineated: we could hold to a maximal Robbian formalization in which Robb's axioms for causal connectibility are retained intact, as are his definitions. We would then have to abandon the postulates which would supplement Robb's original theory and tell us how free particles, rods, and clocks behave relative to the spacetime structures as defined by Robb.

We could adopt an alternative formalization in which the usual connection between free particles, rods, and clocks and spacetime structures is retained, at the same time retaining Robb's axioms for causal connectibility but dropping his definitions of inertial line and of spatial and temporal congruence. We could maintain that this was merely a different formalization of the same theory as the formalization first described, but that it expressed this theory in a different, and preferable, way of speaking.

We could adopt a formalization in which all the consequences of the maximal Robbian formalization were retained intact, and in which the same status—primitive or defined—was attributed to the terms as in the maximal Robbian formalization, but in which the consequences designated "definitions" and "postulates" differed from those so designated by the maximal Robbian formalization. We could maintain that this was a formalization of the same theory as that formalized by the maximal Robbian formalization, but argue that it formalized the theory in a way that gave "more usual meanings" to some of the defined terms. Or we could adopt a formalization exactly like the maximal Robbian's except that in it we refuse altogether to assign "in the margins" definitional or postulatory status to any of the consequences.

Another alternative would be to adopt a formalization like the one described two paragraphs above and declare that this was the formalization of a theory different from that first described. For, we could maintain, Robb was wrong in thinking that his "definitions" of such things as inertial line and congruent interval really were definitions; they could be viewed as empirically testable consequences of the theory, found false by observation.

We could adopt the second kind of formalization discussed above and, again, maintain that it formalized a different theory than that formalized by the maximal Robbian. We could admit that there was no observational means of telling whether the maximal Robbian or this alternative theory was correct, but go on to declare that there could be genuinely incompatible theories that were observationally indistinguishable. We would supplement this approach by offering, one hopes, an account of how a term could have a meaning in a theory without having "directly observational" criteria of applicability, and an account of how one could rationally go about deciding just which of two incompatible but observationally indiscriminable theories was correct. In any case we would either provide this last supplementary account or admit ourselves skeptics or conventionalists with regard to theories.

Finally, we could take the view that all the consequences of the maximal Robbian formalization were correct, but that the maximal Robbian formalization was misleading in that, by incorrectly restricting primitive status to causal connectibility, it misrepresented some of its true observational or factual consequences as having mere definitional status.

V ALTERNATIVE FORMALIZATIONS OF MINKOWSKI SPACETIME

The Formalizations

We have subjected Robb's theory to various stresses, and we have seen how it holds up. When the stress is so great that his postulates for causal connectibility break down, all would agree, I believe, that his metric definitions must go by the board as well. But, as we have seen, a weaker stress, insufficient to contradict any of his axioms for causal connectibility, but sufficient to place stress on the additional postulates connecting the behavior of material objects to the spacetime structure, can lead to many options. Some of these are maximal Robbian, retaining his metric definitions intact; but others lead us to reject the Robbian formalization of a spacetime theory to a greater or lesser degree. Even when the Robbian theory is subject to no observational stress whatever, there may still be dispute as to whether the Robbian formalization of spacetime is correct, and, if "correct," whether it is "best."

Robb's theory is now usually thought of as one way of formalizing special relativity, even if Robb would not quite have viewed it that way. Let us suppose we believe special relativity to be correct. Then we believe that Robb's basic axioms of causal connectibility hold, and that the matter postulates we would ordinarily use to supplement Robb's theory hold as well; i.e., rigid rods correctly measure spatial congruences as defined by Robb, clocks correctly determine Robbian temporal congruences, and free particles follow Robbian inertial lines. If we do believe this, are we then precluded from adopting any of the options discussed in the sections above? We shall see that all of these options are still available to us, each option differing from the Robbian axiomatization as before. But some of the options will no longer have the "plausibility" they had before, in the sense that the grounds for adopting them will no longer be available. On the other hand, some options, previously unmotivated and unlikely ever to be espoused, will now become plausible alternatives.

As before, Option A is the Robbian. The only primitive term is 'causally connectible,' and the axioms for it are Robb's. The metric notions and the notion of inertial line are defined as in Robb. For our complete formalization Robb's theory is supplemented with matter postulates that tell us that rigid rods, clocks, and free particles have the usual special relativistic connection with the underlying spacetime features.

In Option B we would still take causal connectibility as our only primitive notion, and the class of consequences of the formalization we take as primitive would remain the same. But we would offer different definitions for the metric notions and for the notion of inertial line. What would this amount to? Just as our maximal Robbian option amounted earlier to adopting a flat spacetime formalization of a theory which, in general relativistic terms, we would take to be a theory of a nonflat spacetime, adopting Option B would amount to proposing a curved spacetime formalization of what we would take, from a general relativistic point of view, to be a theory of a flat spacetime.

Earlier we saw that one might choose Option B on the grounds that it was "descriptively simpler" than the maximal Robbian formalization, fitting in with the way we would talk in the general relativistic context, as the maximal Robbian formalization did not. But in this case it is clear that the Robbian correlations of the behavior of matter with the spacetime structure are, from the point of view of general relativity, the right correlations; and that if either formalization is descriptively simpler it is the Robbian.

Option C remains quite viable. Here we adopt a formalization that agrees with Robb's in counting only causal connectibility as primitive. It has all the same consequences as Robb's formalization. But it either (a) allocates definitional and postulational status among these consequences in a manner different from Robb's, or (b) eschews the marginal annotation of postulational and definitional status to propositions containing nonprimitive terms altogether.

A proponent of alternative (a) might argue like this: "While all the consequences of the Robbian formalization are true, his allocation of definitional and postulational status to them is untrue to the *meanings* of the 'defined' terms. It is true that free particles travel inertial lines. But that is what we *mean* by 'inertial line.' It is also true that inertial lines have the association with causal connectibility which the Robb formalization postulates. But that is a *fact* about the world, not a proposition true by the meaning of 'inertial line.' So Robb's consequences are all true, but he takes some to be true as a matter of fact that are really true by definition, and *vice versa.*"

The exponent of alternative (a) has a notion of meaning — or at least thinks he does — that allows him to tell what is *really* a definition and what is really a postulate. The exponent of alternative (b) thinks that Robb and the exponent of alternative (a) are equally confused. For, he alleges, there is simply no good reason to label consequences postulational or definitional at all. Remember that he is still allowing that we can determine which consequences of a formalization are "observational," and he is taking difference among these consequences to indicate clear divergence of theory. But beyond that, he believes it is fatuous to pretend that the propositions of the theory involving nonprimitive terms can in any important way be characterized as definitional or postulational. Insofar as the nonprimitive terms are "defined" by the theory, he says, the defining is spread throughout the theory as a whole.

Option D is again one that no one is very likely to espouse. Like Option B it offers a formalization that differs from Robb's in some of the nonprimitive consequences. Thus, like Option B, it amounts to adopting a curved spacetime formalization for what we would ordinarily take to be a theory of a flat spacetime world. Now whereas Option B took this to be a matter of merely "rewording" Robb's theory, Option D takes the new formalization to be that of a theory incompatible with Robb's. Choosing Option D amounts to deciding that spacetime *really* is curved. In one alternative Option D claims that this is an observationally determinable fact; in another, that, while the fact that the spacetime is curved rather than flat is immune to observational discovery, it is still a factual and by no means a definitional matter.

Earlier we saw that this option was a natural one to take in the case in which spacetime was, from the general relativistic point of view, curved but globally conformal to Minowski spacetime. We saw that considerations of the general theoretical context of general relativity might very well lead one to say that the maximal Robbian way of formalizing the theory of spacetime was not just misleading with respect to meanings, but simply *wrong*. And if the maximal Robbian would argue that the two formalizations did not really differ in primitive consequences, the exponent of Option D argued that they did; for he argued that the maximal Robbian's version of what was primitive was too narrow.

But now, given that we belive that the world is, from the general relativistic point of view, flat, who would be likely to argue for a curved spacetime as a theory incompatible with the flat spacetime theory, but correct? If one were to believe that Option D and the Robbian formalization really are formalizations of incompatible theories, he would be more than likely to declare the Robbian alternative correct, and on the same grounds on which the proponent of Option D argued earlier for the correctness of his theory as opposed to that of the maximal Robbian.

The proponent of Option E argues for a formalization that differs from the Robbian in no consequences. But he disagrees with Robb about the scope of primitive concepts. Let us suppose, as we did earlier, that he differs from Robb only in wanting more concepts as primitive, and let us again postpone arguments about the primitiveness of causal connectibility. The proponent of Option E would then argue that, whereas Robb's formalization is based on a correct theory of spacetime, it is misleading in that it falsely declares some of its consequences to be nonobservational or nonfactual. He would say: "I, like Robb, prefer the consequences of Option A to those of Option B. But whereas the Robbian and the proponent of Option B agree that their two formalizations are based on the same theory, I take it that they formalize incompatible theories. So the choice of the Robbian formalization over that of Option B is one of fact, not of mere expression. Like the proponent of Option D, I consider it a factual matter whether spacetime is or is not flat. But whereas he opts for an implausible curved spacetime picture, I opt for the flat."

Can Definitions Be "Right" or "Wrong"?

Let us look more closely at the first alternative version of Option C and at Option E. Both of them find the Robbian formalization defective. The first version of Option C finds Robb's formalization unjust to the meanings of words. Option E finds Robb's formalization unfair to the question of what is "factual" as opposed to "definitional."

Consider the following argument: "In the Robbian formalization the notion of congruence for spatial and temporal intervals is *defined* by means of causal connectibility. So the connection between the notion of congruence and the causal notion used to define it can hardly be said to be correct or incorrect. These, like all definitions, in order to be acceptable, must meet certain formal requirements of existence and uniqueness of defined entity. Beyond this, there is no sense in speaking of a definition as 'right' or 'wrong.'"

But the Robbian definitions of spatial and temporal congruence are not the only possibilities for causal definitions of "congruence" relations in his theory. One can, as Malament has shown, causally define, in Robbian fashion, a congruence relation which "renders as congruent timelike and spacelike vectors whose Minkowski lengths are equal in absolute value (although opposite in sign). And one can define a relation that renders timelike and spacelike vectors congruent if the absolute value of the Minkowski length of the former is p/q times that of the Minkowski length of the other, where p/q is any particular positive rational number. Both these definable congruence relations otherwise agree with the standard congruence relation."[18] Of course, if we so define 'is congruent to,' rods and clocks will not, even in Minkowski spacetime, determine congruity correctly when used in the usual way. But why should that bother us if the only meaning attribution we have for 'is congruent to' is our causal definition?

But these definitions of the congruence relation would be *wrong*. They would not be causal "definitions" of 'congruent,' but of some other relationship misleadingly so named. This shows that, in some sense at least, these "definitions" are not really the appropriate definitions.

Here I think the argument could go in one of two directions, one appropriate to the advocate of the first version of Option C and the other appropriate to the advocate of some version of Option E. In

18. D. Malament, private communication. These results are proved in Malament's doctoral dissertation for Rockefeller University, "Some Problems concerning the Causal Structure of Space-Time." Also discussed in that thesis are a number of distinct notions of "definability" slurred over, I hope without philosophical loss, in this paper.

the first version it is agreed that 'is congruent to' can only be a defined term in a formalization. But, this version argues, it is wrong to the *meaning* of 'is congruent to' to give it a *causal definition*. For example, one might argue that the association of congruence with rigid rods and clocks really captures the meaning of these terms. One might argue that the usual definition of congruence in terms of rigid rods cannot be "wrong" or "right," for it is this association that really fixes the meaning of 'is congruent to' as we usually use that expression. Congruent intervals are *just* those intervals determined congruent by the usual method of transported rods and clocks. What masquerade as definitions in the Robbian formalization are really postulates that associate certain causally definable structures with the congruence relation. The association will indeed hold if spacetime is Minkowskian, but one hardly wants every general truth about the world to count as a "definition."

A similar but more complex tale about "meaning," or perhaps a denial that 'meaning' has any useful meaning and that out with it goes any useful notion of 'definition' in a formalization of this kind, might accompany the view that we should either talk about the meaning of 'is congruent to' as being fixed by its total role in the theory, or not talk about its meaning at all. In any case this position would still reject the claim that the causal definitions correctly "fix the meaning" of 'is congruent to' in any sense beyond correctly designating as congruent those intervals that are in fact congruent when the spacetime is Minkowskian.

The second version argues that 'is congruent to' should be taken as part of the primitive vocabulary. This signifies that both the postulates connecting causal structure to the congruence relation *and* those associating the behavior of rigid rods and clocks with the congruence relation are empirical postulates that all turn out to be true because of the Minkowskian structure of spacetime. As usual one might argue that these postulates are observationally determinable to be true or false, or that they are factual but not directly observational. But, in any case, in this view none of these postulates constitutes a "definition" of the congruence relation, for that is primitive and undefined. From both these general points of view one can see why we would instinctively feel we can judge Robb's definitions to be "right" or "wrong" in a manner entirely inappropriate for definitions.

We noted earlier that in Minkowski spacetime (and in a more

general class of spacetimes as well), we could "define" a topology
—the Alexandroff topology—that would agree with the usual
manifold topology on the spacetime. We might note here (and we
shall return to this point later) that what we have said about causal
definitions of metric notions holds for causal definitions of topological notions as well.

Zeeman, for example, has shown how one can, in Minkowski
spacetime, causally define (if one allows oneself the full resources of
set-theoretic definition) a topology quite different from the Alexandroff topology, and hence from the usual manifold topology.
The Alexandroff topology is the coarsest topology in which the
interiors of light cones are all open sets. The Zeeman topology is the
finest topology on the spacetime that induces the usual three-dimensional Euclidean topology on every space slice of the spacetime
and the usual one-dimensional Euclidean topology on every time
axis.

This topology has a number of curious features. For example,
the topology induced on light rays by it is discrete. In this topology
the paths of freely moving particles are all piecewise linear.
Interestingly—and perhaps not too surprisingly, in the light of
Zeeman's other work mentioned above—every one-to-one mapping of a Minkowski spacetime onto itself which is bicontinuous in
this topology is first of all a causal automorphism (i.e., it preserves
all causal connectibility and symmetric absolute temporal order
relations); and is, second, one of the group of mappings generated
by inhomogenous Lorentz transformations and dilations![19]

Thus, just as many "congruence relations" are causally definable, so are many "topologies." Later I shall deal to some extent
with the question of how to pick the right topological definition. I
believe the discussion above, however, in the light of such novel
"causal topologies" as Zeeman's, is enough to show us that one can
no more glibly assume there is no question of rightness or wrongness about a causal definition of the topology, than one can assume
there are no such questions about the causal definitions of the
metric structure. And if there is a question of rightness and wrongness, in what sense can these "definitions" be said to be "merely
definitions"?

19. E. Zeeman, "The Topology of Minkowski Space," *Topology* 6 (1967): 161–
170. I am indebted to David Malament for pointing out to me the relevance of this
work in the present context.

VI ON DRAWING PHILOSOPHICAL CONSEQUENCES FROM FORMALIZATIONS

Winnie's Thesis

At long last we can consider one example of an answer to our second overall question. So far we have been asking: what are the possible alternative formalizations of a theory, and how do they differ from one another? We have pursued in some detail our first primary question: what could be the scientific and/or philosophical grounds for preferring one formalization to another? Now we wish to ask: what are the philosophical consequences of adopting a given formalization? Since we see that much philosophical decision-making goes into choosing a formalization, we shall really be exploring the question: how do the philosophical choices one makes in choosing a formalization affect further philosophical consequences?

The philosophical theses Winnie extracts from Robb's formalization of Minkowski spacetime provide, as the reader will recall, an example of consequences drawn from formalization. Winnie's argument went like this: Robb has shown us that in the case of Minkowski spacetime the metric is reducible to the topology. But some authors have claimed that it is because the metric is undetermined by the topology that the metric is a matter of convention. So, in the case of Minkowski spacetime, we see that the metric is not conventional, at least not in this sense.

There are two issues we must pursue:

(1) In what sense, or senses, is it true that in Minkowski spacetime the metric is "reducible" to the topology? In what senses is this false?

(2) Is the reducibility, in any sense, of the metric to the topology ever good grounds for declaring the metric nonconventional? Is nonreducibility of the metric to the topology, in any sense of 'reducible', ever good reason for declaring the metric conventional?

On Causal Theories of the Metric

Robb has shown, beyond question, a number of interesting and important things. He has shown that if spacetime is Minkowskian, then the fundamental metric relations of congruence, and the property of being an inertial line, are coextensive with relations and properties definable purely in terms of the absolute "after" relation — or, if you will, in terms of causal connectibility. But has he shown that, if spacetime is Minkowskian, the metric is "reducible" to the topology?

Using the results noted earlier concerning the causal definability of the Alexandroff topology and its extensional equivalence with the manifold topology when spacetime is Minkowskian, together with Robb's results, the following can be derived: if spacetime is Minkowskian, we can determine all the topological properties of regions of the spacetime simply by determining which events are and which are not causally connectible. Using the very same determinations, and no others, we can also find all the metric properties of regions of the spacetime.

But this hardly means that we can, by using causal connectibility alone, determine *if* spacetime is Minkowskian, knowing only that it is pseudo-Riemannian. We can tell that it is not Minkowskian if one of Robb's postulates fails; but even if none does fail, this is no guarantee that the spacetime is flat. And the correctness and importance of Robb's results within his formalization hardly force us to accept the view, even if we take it that spacetime is Minkowskian, that he is right in saying that causal connectibility is an immediately determinable fact in a theory-independent way, and hence a proper member of the class of primitive terms. Nor does Robb's formalization mean we must agree that he is right in consigning all the metric notions to the nonprimitive class on the basis of "nonobservability." Even if we do partition the terms as he suggests, we need not accept his "definitions" as truly definitional, nor agree that the *meanings* of the metric terms are correctly fixed by their causal "definitions." As we have seen, there are generally good reasons for adopting just such "anti-Robbian" positions. And, I submit, if we consider just how little we must accept, given Robb's results, and just how weak the connection is between causal order and metric structure, compared with what philosophers have usually demanded of an "X theory of Y," we shall probably conclude that Robb's results certainly do not establish a "causal theory of the metric," nor a "reducibility of the metric to causal order" in many of the most important philosophical senses of 'reducible.'[20]

20. For a discussion of the philosophical issues to follow in a slightly different scientific context, see Sklar, *Space, Time, and Spacetime,* 318–343.

To espouse an "X theory of Y" might be to maintain that X and Y are coextensive predicates. Now we do not believe that metric features will in fact be coextensive with the Robbian causal correlates, since we believe that spacetime is not Minkowskian. But if we believed that spacetime were Minkowskian, I think we would have to agree that a causal theory of the metric was correct in this very weak sense. Alternatively, to espouse an "X theory of Y" might be to maintain that X and Y are coextensive as a matter of scientific *law*. Obviously we do not believe that the causal and metric predicates are so lawlike coextensive, since we do not believe they are coextensive at all.

But suppose we believe that spacetime is Minkowskian. Believing, as we do, in general relativity, I think we would still deny lawlike coextensiveness to the pair of predicates. For, from the point of view of general relativity, the coextensiveness is a matter not of the fundamental laws of physics, but of the de facto flatness of spacetime. Minkowski spacetime is just one possible model for the laws of general relativity, but hardly the only "lawlike possible" spacetime. But if we were working in 1905, say, and had never even seriously contemplated the possibility of curved spacetime, the universal (for all of spacetime) coextensiveness of the causal and metric predicates might be taken by us as a fundamental law of nature.

We noted earlier that if we believed in general relativity, observations on causal connectibility alone would not be sufficient to fix for us the curvature of spacetime. This illustrated a fundamental fact: the question of what kinds of observations would or would not be sufficient to determine fully all relevant physical features of the world depends upon what possible worlds one is willing to contemplate. And that depends upon the fundamental theory one believes. Thus if we presupposed that spacetime was Minkowskian, we would agree that we could map its metric using causal connectibility alone. But if we only presupposed spacetime to be pseudo-Riemannian, we could not. Now we see that our presupposition about the possible worlds, framed on the basis of our most fundamental theory, has other consequences. For whether a correlation of features is taken to be lawlike or merely de facto is again a function of our *assumptions* about which possible worlds we will consider.

To espouse an "X theory of Y" might be to maintain that the predicates are coextensive as a matter of their *meaning;* that 'all Y's are X's' is an analytic truth. In the philosophical contexts we are interested in, the argument for this usually takes a standard form: we can tell what an X is without presupposing the theory in question. But we have no independent check on what a Y is; we simply call Y's the things that are X's. For a causal theory of congruence this would amount to claiming that we can, without presupposing the theory in question, determine which events are and which are not causally connectible. But we have no independent check on the metric. What we mean by 'congruent intervals' is some relationship between two pairs of points characterizable in terms of causal connectibility.

Believing in general relativity, we do not accept this argument. For we believe that there are nonisometric spacetimes with all the same causal connectibility features between the point-events mappable one-to-one on each other. But suppose we thought spacetime was Minkowskian? Would we believe in such "analytic coextensiveness" then?

Once again our thinking depends, I believe, on a complex estimate of our theoretical situation, not only on the theory we do believe, but on the theories we did but no longer believe, and on the theories we think we might believe should observation ever lead us to give up our present theory. Our views about the formalization of Minkowski spacetime are nice test cases for us. For this theory is the successor to previous spacetime theories, and has itself been succeeded by a more general theory. This case will, we shall see, be applicable in interesting ways to the case in which we have no specific "future theory" yet in mind. For example, it will be useful when we come to discuss the "meanings" of terms in formalizations of general relativity.

Which propositions of a theory are "analytic," and which are true "by the meanings of the words alone"? We know we do understand what we are asking here; philosophical positions range from the view that there is something to analyticity but we do not really know what it is, to the view that the whole analytic/synthetic distinction is just a tissue of confusions. But perhaps the following will throw some light on the problem.

The "analytic" propositions of a theory should be at least more or less "fixed points" in theoretical change. For if a proposition is analytic, what on earth could lead us to deny its truth? Obviously the situation is not that simple, for one could always offer a new theory in which the proposition is denied, yet attribute analyticity to it in its old role by arguing that the meanings of the terms have changed in the theoretical transition. After all, we have seen how one can argue for different formalizations of even one and the same theory, which may or may not differ in asserted consequences but which do differ in their "meaning" attributions to defined terms.

Yet I think the following position has some validity: at least a prima facie guide to the analyticity of a proposition in a theory—if there is such a thing—is the stability of that proposition in the transition from an older theory to the one now believed, and from the one believed to one of the ones we presently contemplate as theoretical alternatives to which we would move under the pressure of novel observations.

For example, in Newtonian spacetime theory we maintain that rigid rods measure spatial intervals. They do so as well in Minkowski spacetime understood from the special relativistic point of view. And in general relativity we still say that rigid rods correctly determine spatial congruence. So it may be plausible to argue that the connection between rigid rods and spatial congruences in Minkowski spacetime theory is "analytic," for it is that connection that remained fixed when we moved from the Newtonian to the Minkowskian theory, and remains fixed when we move from Minkowski spacetime to general relativity and curved spacetimes.

On the other hand, Robb's association of spatial congruence with causal connectibility holds only in the Minkowskian theory, and not in the Newtonian or in the general relativistic case. Isn't that some reason for attributing analyticity to the rigid rod-congruence connection in the theory of Minkowski spacetime, and for denying analyticity to the Robbian association of congruence with causal connectibility? What other, or better, reason could there be for distributing analyticity or nonanalyticity than this "historical" fixedness or nonfixedness of the connection?

Once again we see that our philosophical attribution will depend on many things. First of all, it will depend on assumptions about the independent determinability of various features of the world. And it will depend not only on the relation of the theory we are interested in to the theory which preceded it, but on its relation to the class of theories that we *assume* to be plausible candidates for a novel theory to adopt under observational pressure which would lead us to relinquish our present views.

To espouse an "*X* theory of *Y*" might mean to maintain that the coextensiveness of the predicates was somehow a *necessary* truth, not identifying this notion of necessity with either lawlikeness or analyticity. Despite recent invocations of possible world semantics and doctrines concerning necessary truths about individuals or substances which are neither a priori nor, perhaps, analytic, I cannot understand what such an allegation would amount to. Once again the clearest notion I can think of relies upon putting the theory in question in the framework of antecedent theories and of assumed alternatives. Perhaps it would be plausible to speak of some proposition of a theory as necessary if it held not only in this theory, but in all antecedent theories and in any plausible alternative to which we would turn if we abandoned our present theory in the light of new data.

Under this interpretation the causal theory of the metric in Minkowski spacetime again seems false. For although the congruence relations are coextensive with certain causally definable relations in Minkowski spacetime, they are not so coextensive in either the theory that antedates Minkowski spacetime or in the general relativistic theories to which we would turn, even if we believed in Minkowski spacetime, should the data indicate to us that something about the special relativistic picture of the world was wrong in its assumption of flatness of the spacetime.

The reader may object that my notions of the lawlikeness, analyticity, or necessity of the connection between the relations were all defined only relative to a body of assumed theoretical alternatives to the believed theory. I reply that this is the major virtue of my analysis of these notions. Once again the claim is that while *facts* and *conventions* have received their due in the philosophical theory of the epistemology, semantics, and metaphysics of theories, *assumptions*—assumptions about the class of theories we will even consider as possible candidates for belief—have not.

Can One Get from Causal Definability to Nonconventionality?

I have argued that a causal theory of congruence is plausible, even if we believe in Minkowski spacetime, only in its very weakest sense —the sense of mere coextensiveness in this spacetime of some congruence relations and some relations definable by causal connectivity. But what about the claims that are made for what this shows concerning the relation of metric to topological structure in Minkowski spacetime? And what about the claim that this "causal theory of the metric" is relevant to showing the metric "nonconventional?"

First we should note that what is offered in the Robbian formalization, even in the weakest sense of an "X theory of Y," is a theory of congruence in terms of causal connectibility. How then is congruence associated with *topology* in this view? The connection is mediated by the allegation that one can offer a theory of the topology in terms of causal connectibility as well. We do not offer topological "definitions" of metric quantities; rather, we attempt to "define" both metric and topological notions by causal ones.

Later I shall argue that the causal "definition" of the topology in Minkowski spacetime is a "definition" only in the very weak sense in which we have seen it possible to "define" the metric causally. That is, I shall argue that while it is possible to construct, out of causal connectibility, properties and relations coextensive in Minkowski spacetime with all the relevant topological properties and relations, there is good reason to think these coextensivities are neither lawlike, analytic, nor necessary—just as was the case with the causal and metric coextensivities.

So let us speak of a *causal* theory of the metric in Minkowski spacetime, rather than of a *topological* theory of the metric. Is there any reason to think that the metric of Minkowski spacetime is "nonconventional" because it is *causally* "definable?" I think one could argue from "causal definability" to "nonconventionality" only in the following way: which events are and which are not causally connectible is a fact discernible totally independently of any theoretical assumptions. Hence no elements of "convention" enter into deciding which events are or are not so related. But there is, in addition, no "optionality" or "choice" in going from causally connectible events to the metric of the spacetime. Therefore the "nonconventionality" of the metric follows from its reducibility to causal connectibility.

Since other arguments have been used to associate nonconventionality with "reducibility to the topology," we might first say something about them. Riemann suggests, and Grünbaum has held, the view that if the topology of spacetime were discrete, we could determine metric relations from topological by "counting numbers of points between events." This is mysterious in the case of multiple dimensional spacetimes since we do not yet understand, I believe, what such discrete multidimensional cases would amount to. Even in the one-dimensional case of a simple ordered set, it is not clear why we should take the *real* distance between points to be a function of the number of points between them.

Further, even if the metric were determined by counting points, it is still hard to see the relevance of this to issues of conventionality without some demonstration, never given, as to why counting points is not in any way conventional. In any case, this is hardly relevant with respect to Minkowski spacetime, where the topology is admittedly dense and not discrete.[21]

Again, it has been argued that since many metrics are "compatible" with a given topology, this is why the metric is "conventional." But if "compatible" means only that both the metric and the topology are consistently assignable to the underlying point set of events, then, trivially, many topologies are "compatible" with a given metric. Zeeman's results show that there are even many topologies compatible with the Minkowski metric in the further sense that the metric is continuous with respect to each of them. In any case, the fact that many "metrics" in the formal sense are "compatible" with a given topology hardly shows that the real metric — the real measure of intervals between points — is a matter of "convention" in any interesting sense. And why should we, without further *epistemic* argument, take the topology as somehow "less conventional" than the metric?

We have seen many reasons for thinking that even given a "non-conventional" notion of causal connectibility, we should be dubious about the move from that to a "nonconventional" metric. In many spacetimes no "definition" of the metric congruence in terms of causal connectibility is even possible. Even when one would be possible (i.e., when the Robbian postulates for causal connectibility hold), we might be reluctant to accept these "definitions" as even extensionally correct (the case of spacetimes globally conformal to Minkowski spacetime but not flat). And even when we admit the coextensivity to hold, we have seen good reasons for denying that even then the "definitions" of the metric in terms of causal connectibility are really definitions in any deeper sense than that of true statements of coextensiveness. And if that is all the definitions are, it is hard to see how they provide an adequate bridge from the nonconventionality of the causal connectibility relation to that of the metric. Couldn't we still "conventionally choose" a different metric, simply by paying the price of denying the coextensiveness of the metric and causal connectibility relations, on the

21. Sklar, *Space, Time, and Spacetime,* 109–112.

grounds that they were not, after all, in any sense "necessary" or even "lawlike"?

Even if we could establish the causal connectibility relations so as to preclude any invocation of conventionality in the process, and even if the relations so established allowed for some "definition" of the metric in terms of causal connectibility, what is there to prevent us from asserting that accepting those "definitions" is "just a matter of convention"?

But is even the establishment of the causal connectibility relations a matter totally immune from "conventional" elements? Let us look at two versions of theories of how causal connectibility relations are established, and see how their "nonconventionality" fares under each. One version is that espoused by Robb and, as remarked, interestingly like the position taken by Bergson. Let us do the analysis in terms of Robb's 'after' relation. Why is it not a matter of convention which events are truly after which others? Well, says Robb, because we could imagine a human consciousness moving from one event to another absolutely after it in time. The 'afterness' relation between the events would be an immediate apprehension by this consciousness, and such immediate apprehensions are matters in which no conventionality can rear its ugly head.

Notice, first of all, how we have retreated "behind the veil of perception." Isn't it *subjective afterness* which we are now determining, and not an *objective* physical relationship? The reader should compare this with Reichenbach's remarks, noted above, concerning which notion of coincidence of events should be taken as primitive in formalizing special relativity. In addition, it is hardly alleged that some consciousness does actually record the afterness between any pair of events—only that one could. So the nonconventionality is a matter of *potential* direct apprehendability, rather than *actual* direct apprehendability.

A minor technical difficulty with this Robbian approach also arises. Events that are light-connectible, but not particle-connectible, are still in the absolute 'after' relationship. But what sense does it make to talk of even a possible consciousness experiencing directly the "afterness" between two such light-connectible events, given the embodiment of consciousness in material bodies and the impossibility that such bodies, even in principle, could travel at the velocity of light? Some "limit of possible experience of a single consciousness" might salvage Robb's idea; or he could take as primitive, instead of 'after,' the notion of 'connectible by a genidentical *material* signal.' But I shall not pursue this here. It is interesting to note that in Woodhouse's formalization, which we shall later examine, continuous timelike paths are taken as primitive, but continuous light paths are not.[22]

Another item of philosophical interest arises if we imagine a plurality of consciousnesses, one for each possible timelike worldline, "checking up" on afternesses. Would an actual or potential *social* direct apprehendability do, or must we imagine—contrary not only to fact but to lawlike possibility—a *single* "direct apprehender" to check up on which events are really, and nonconventionally, after which? Here the reader may want to reflect on the relation of this issue to the so-called problem of indistinguishable spacetimes. This shows that there can be spacetimes that no single observer could ever tell apart on the basis of his experience, even though we postulate them to have quite different structures. This is because of the causal limits on the information a single observer can receive from or communicate to other observers. If we allow "potential social awareness," an imagined plurality of observers who cannot, in fact, pool their information, but whose total knowledge *we* can imagine in toto, then such universes are "distinguishable" by the "assembled observational data." The notion of direct apprehendability, then, is far from clear. And even if it were clear, it is hard to see how one could get from the subjective realm of appearances to a *distinct* realm of "nonconventional" objective relationships.

The other approach goes like this: let us now work in terms of causal connectibility instead of in terms of 'after.' What makes causal connectibility "nonconventional" is the direct empirical test for it in terms of genidentical particles or light waves. Two events are causally connectible if and only if a possible genidentical signal can be sent from one to the other.

One difficulty with this approach is its "scientific unreality." Given that we really believe in quantum mechanics, and not in the classical pseudophysics which usually serves as the arena for foundational discussions in relativity, to what extent do we really want to pin down our foundational expositions to those that rely on such

22. N. Woodhouse, "The Differentiable and Causal Structure of Space-Time," *Journal of Mathematical Physics* 14 (1973): 495–501.

concepts as single genidentical particles and their paths, when, on the basis of our real physics, we are in some doubt about the very existence of such classical objects? But let this be for the moment.

A more direct philosophical criticism is this: we are trying to establish a "nonconventionality" for causal connectibility. To do this we are using the notion of being connectible by the path of a genidentical particle or light wave. But is it obvious that the question of which events are so connectible is a matter totally devoid of "conventional" elements?

Reichenbach, for example, considered the very topology of spacetime as conventional as its metric, in the sense that we could choose, "conventionally," different topologies for spacetime and still "save the phenomena" as long as we made sufficient changes elsewhere in our physical theory—in particular, changes in what we took to be paths of genidentical particles or light rays. One such choice would depend upon how we identified or "disidentified" events. This is what happens when we "conventionally" choose between a spacetime and a different spacetime that is a covering space for it. But, Reichenbach thought, even the *local* topology was conventional, for we could "conventionally" select spacetimes with different "nearness" or "neighborhood" relations by deciding differentially on which paths were or were not continuous. Now Reichenbach thought that such "conventional" choices were only choices about "manners of speaking" or "descriptive simplicity"; but we have seen that other kinds of conventionalism could make use of the same arguments. For example, we could espouse the conventionalism that tells us that the Reichenbachian topological alternatives are *really* incompatible theories among which we could make no choice on observational grounds.[23]

So if we try to argue the nonconventionality of causal connectibility on the grounds of its "empirical testability" by genidentical particles, we must first of all make the usual philosophical moves of allowing possible genidentical particle connectedness to test causal connectibility, and, more important, have some further argument, *contra* Reichenbach, to the effect that connectibility by a genidentical particle was itself "nonconventional."

I shall argue later that there are some grounds for making this move. That is, I shall give some reasons for maintaining that gen-

23. Reichenbach, *Philosophy of Space and Time,* section 12.

identical paths are nonconventional. I shall base this view on arguments for defining genidentical paths in terms of continuous paths, rather than vice versa; and on arguments for taking one kind of topological nearness to be as primitive as directly observationally determinable—in a theory-independent way—as anything can be in a foundational analysis of general relativity. So I am not saying that causal connectibility isn't correctly designated 'primitive.' I am only saying that if one makes the decision to so consider it, one must argue for this decision in a complex philosophical way; one should not just assume it.

My general response to Winnie's argument from Robb's formalization to the nonconventionality of the metric in the theory of Minkowski spacetime is similar. One can, perhaps, draw philosophical consequences from the existence of certain formalizations of scientific theories. But one can do so only after one has chosen the formalization. In doing so we are operating on the basis of numerous, complex, and sophisticated *philosophical* presuppositions. Even having chosen a formalization, further *philosophical* assumptions may be necessary to get the philosophical results out of the formalization. We can get philosophical conclusions out if we put philosophical assumptions in. These assumptions should be clearly laid on the table. And once they are, we see that they are rarely obvious or self-evident, but rather the end product of a complicated chain of philosophical reasoning.

In Winnie's particular example, two grand steps must be taken. First, we must agree that in the theory of Minkowski spacetime, the metric is truly definable, and not just "definable" in terms of causal connectibility. We have seen just how dubious this thesis is. Second, we must agree that causal connectibility itself is a nonconventional matter. While this may be true, it must be argued. And the arguments for it are neither obvious nor simple.

VII FORMALIZATIONS AND GENERAL
RELATIVISTIC SPACETIME

So far we have been studying, primarily, the problem of the formalization of Minkowski spacetime from a philosophical point of view. But we know that spacetime is not Minkowskian. Why should we have gone to all this effort over a dead theory? This answer is clear: the apparatus we have developed will be of great use in trying to understand, philosophically, the numerous problems that arise when we try to formalize the spacetime theory of general relativity, our currently most viable theory of spacetime. It will also be helpful, I believe, in the general methodological problem of formalizations, even outside the area of spacetime theories.

When we come to examine formalizations in general relativity, we should note initially a couple of interesting features of this case that distinguish it from formalization of Minkowski spacetime. First, in formalizing a theory of Minkowski spacetime, we are concerned with the theory of a single spacetime model. In considering formalizations of the spacetime theory of general relativity, we must remember that this one theory allows at least the lawlike possibility, relative to its laws, of many distinct spacetime worlds. So when we wanted to put "stress" on the Minkowski spacetime theory by looking at alternative spacetimes, we had to "go beyond the theory whose formalization was in question." But we can make progress in understanding the philosophical presuppositions of formalizations in general relativity, by considering the wide variety of spacetimes compatible with the theory. This is a variant on the old remark that in special relativity, as in the usual Newtonian theories, spacetime is a "passive arena" of material events. But in general relativity, spacetime is itself a variable dynamic element in the theory.

Second, when considering the philosophical aspects of formalizations in the theory of Minkowski spacetime, we were able to avail ourselves not only of the theories of spacetime which preceded the one we were primarily concerned with, but also of its successor theory, general relativity. We saw that our views about the adequacy and desirability of various formalizations could be radically affected by the fact that we had this newer theory of spacetime in mind. Essentially we could ask how well a given formalization of Minkowski spacetime "fit in" to the future developments in the theory of spacetime whose nature we already knew. In examining the formalizations of general relativistic spacetime theories, we are not in this enviable position. For while alternative theories to general relativity have certainly been espoused, and while some of these are "generalizations" of the general relativistic theory, in the sense that this theory is a generalization of Minkowski spacetime theory, there is no more general theory of spacetime beyond general relativity that is universally agreed to have plausibility either as a correct theory of spacetime (general relativity being assumed incorrect), or even as the theory of spacetime to which we *would* move were we to come to reject the general relativistic theory. We shall see that this is of some importance in our philosophical analysis.

From Rods and Clocks to Clocks: Synge's Chronometric Formalization of General Relativistic Spacetime

Suppose we believe in general relativity. How should we go about determining just which of the many spacetimes compatible with the theory is the actual spacetime of the world? And how should we determine the actual intervals between events within this spacetime world?

The natural suggestion is to use infinitesimal rigid rods ("taut strings," if you wish) to measure spacelike separations in some observer's frame, and to use ideal clocks to measure timelike intervals. From these the general intervals between events can be determined; with enough such measurements performed on a sufficiently fine scale, we can determine the g-function at every point of the manifold and hence both the structure of the spacetime and the particular intervals between particular events.

Now many physicists have been reluctant to accept rigid rods, even infinitesimally, as "primitives" in their formalizations of general relativity. Others wish to eschew as well the use of the notion of ideal clocks as a primitive. For the moment I want to avoid discussing their reasons for this self-denial. I shall, however, look closely at this issue in a later section.

Synge is reluctant to use infinitesimal rigid rods, but he is perfectly happy with clocks. And he has shown that using clocks alone as primitive we can formalize the theory.[24] How does he proceed? First, he takes as primitive a class of ideal clocks. Their rates are all linearly related to one another, in the sense that two of these clocks transported together from one event to another along the same timelike world-line will have their number of "ticks" elapsed between the two events in a constant numerical ratio.

Next he makes the following assumption: suppose x and $x + dx$ are two events connectible by a timelike world-line. Let ds be their interval separation. Then $ds = f(x, dx)$, where f is positive homogeneous of degree one in the differentials. Now it has been alleged that

24. J. Synge, *Relativity: The General Theory* (Amsterdam: North-Holland, 1960), chap. 3, especially sections 1 and 2.

he offers no justification for making such an assumption.[25] But the justification which, I presume, he is presupposing is that offered by Riemann for assuming at least this much about the metric form in his original inaugural dissertation on Riemannian geometry.[26] First we must assume that the g-function will vary continuously in the coordinates to show that ds is a function only of x and dx. Next, if we assume that, ignoring magnitudes of second and higher order, the length of a line element remains unchanged when all its elements undergo the same infinitesimal change of place, then the homogeneity of f in the first order in the differentials follows.

But for general relativity we need much more than this. We need the usual *quadratic* form for ds as a function of the dx's. Synge assumes this form for the g-function, and here he does so with no further rationalization of the assumption. Once we have made this assumption, however, Synge shows how, by making a sufficient number of interval measurements between timelike connectible events, using his ideal clocks, we can calculate the g-function at any point. But if we have the g-function, we can determine the lightlike and spacelike separations between all point-events as well. Thus using ideal clocks as our only primitives, assuming them to measure timelike intervals correctly, and assuming, boldly, the full quadratic form of the g-function, we can determine the metric structure of our spacetime fully, and we can determine the particular interval separations between all events, even those lightlike or spacelike separated.

Causal-Inertial Theories of the Metric

But there are those who will, for reasons we shall examine later, eschew *clocks* as primitives along with rods. What will they take as primitives? In a "rods and clocks" formalization, or in a "clocks alone" formalization, we shall need to supplement our initial postulates about the adequacy of the primitive devices as interval measurers, and our initial assumptions about the form of the g-function, by postulates that relate the behavior of light rays in a vacuum and free particles to the structure of spacetime. Usually we

25. A. Grünbaum, "Geometrodynamics and Ontology," *Journal of Philosophy* 70 (1973): 775–780; see p. 780. I am indebted to this article by Grünbaum for suggesting to me many of the crucial questions discussed in this essay.
26. B. Riemann, "On the Hypotheses Which Lie at the Foundations of Geometry," in D. Smith, *A Source Book in Mathematics* (New York: McGraw-Hill, 1929), 411–425. The relevant passages are on p. 416.

shall simply assume that light rays travel null geodesics, and free particles, timelike geodesics. Perhaps we can get a different formalization of spacetime theory by taking these entities as the primitives of our theory. Since light rays in a vacuum mark out the boundaries of the sets of causally connectible events, we shall be searching for a formalization in general relativity that is as close as we can come to Robb's formalization in Minkowski spacetime theory.

Since we know that nonisometric pseudo-Riemannian spacetimes are mappable one-to-one, preserving all causal connectibility relations, we shall not attempt a formalization that takes light rays alone as primitive. This was Weyl's mistake, as Lorentz pointed out. But the following fact gives us reason to believe light rays and free particles together will be enough: any two pseudo-Riemannian spacetimes that are causally isomorphic under an isomorphism which takes timelike geodesics into timelike geodesics will be isometric, up to a constant factor. The fact that, in general relativity, free particles travel timelike geodesics (either because we introduce this as a separate postulate or because we show that they must so travel because of the nonlinearity of the field equations), suggests that light rays, marking out the causally connectible from the non-causally connectible, and free particles, marking out the timelike geodesics, will be enough. Actually, we must be careful here, as we shall later see. For by 'free' particle we mean more than is apparent at first glance. Free, infinitesimal, gravitational monopole would be more correct. But I shall deal with this later.

Several methods for measuring the spacetime using light rays and free particles alone have been outlined. The interested reader should consult Marzke and Wheeler, and Kundt and Hoffman, for details.[27] The formalization I shall focus on, which is a generalization of the Robbian approach, is the extremely interesting and important one of Ehlers, Pirani, and Schild, which I shall call the EPS formalization.[28]

To understand the EPS formalization we shall first have to look

27. R. Marzke and J. Wheeler, "Gravitation as Geometry — I: The Geometry of Spacetime and the Geometrodynamical Standard Meter," in H. Chiu and W. Hoffman, eds., *Gravitation and Relativity* (New York: W. A. Benjamin, 1964). W. Kundt and B. Hoffman, "Determination of Gravitational Standard Time," in *Recent Developments in General Relativity* (New York: Pergamon, 1962).
28. Ehlers, Pirani, and Schild, "The Geometry of Free Fall," 63–84.

at some of the mathematical aspects of general relativistic space-times. The full pseudo-Riemannian metric manifold of general relativity is a complex structure. From it we can abstract some simpler structures.

First of all, the manifold is a topological space. Second, it is a differential manifold. Next, these spacetimes have a *conformal structure.* Infinitesimally this means the possibility of constructing the hyperplanes orthogonal to a given line element. Such a construction results in the infinitesimal null-cone structure, i.e., in the distinctions among timelike, null, and spacelike connected events, or, physically, in the causal connectibility structure on the space-time.

The spacetimes also have a *projective structure.* Infinitesimally this means that there is a well-defined notion of parallel-transport-ing, in its own direction, a vector from tangent space to tangent space. Globally this results in a geodesic structure on the spacetime. For the geodesics are just those curves whose tangent vectors re-main tangent vectors when they are parallel-transported in their own directions.

The spacetimes have an *affine structure.* Infinitesimally this means that there is a well-defined notion of parallel-transporting, in any direction, a vector from tangent space to tangent space. Globally this results in the definability of an affine parameter along the geodesics. Equidistant points along a geodesic (with respect to the affine parameter) are those points whose "connection vectors" are parallel.

Finally, there is the full metric structure, which allows us to compare the intervals between arbitrary pairs of pairs of point events.

EPS assumes the notions of event, light ray, and particle. Parti-cles are meant, intuitively, to be paths of "free" particles. Axioms are introduced concerning the smoothness of the particles and of messages sent between them by light rays. An axiom is introduced which allows us to coordinatize the space by particles and light rays, and it is assumed that any two such coordinatizations are smoothly related to each other. This lets us define a differential topology for the manifold of events. There are some problems here, but I shall discuss them later. Next, axioms on the light rays are introduced that are sufficient to allow us to define the usual light-cone struc-ture on the spacetime, i.e., to define its conformal structure. Then axioms on the particles are introduced that are sufficient to let us define the projective structure of the spacetime.

But we still don't have a full affine structure for the spacetime, much less its full metric structure. EPS shows that only one additional axiom is necessary to guarantee the existence of the full affine structure. This is an axiom of compatibility between the conformal and projective structures already introduced. It states that the interiors of the light cones, definable by the conformal structure, are filled with the particles. That is, we can approach the light rays as closely as we like by paths of free particles in the interiors of the light cones. It follows that a full affine structure is definable on the space by the light rays and particles alone.

Finally, we can get the full metric structure in either of two ways. We can demand that if we have two nearby timelike geodesics, and if we mark off equal intervals along them relative to their affine parameters, the events marking off the equal intervals will be simultaneous (to the first order) using the "radar" method for establishing simultaneity familiar from special relativity. Alternatively, we can define congruence for vectors that are located at the same point by using the conformal structure. We can define parallel transport of a vector along a curve as well, using the affine structure: if we take a vector and transport it parallel to itself to a new point by two different paths, then, if the two resulting vectors are congruent at this second point, a full pseudo-Riemannian metric structure on the spacetime is well defined. Thus if we add either of these last two conditions as an axiom, we can fully define the metric structure of the spacetime using light rays (causal connectibility) and particles (timelike geodesics) alone.

Ehlers, Pirani, and Schild claim a number of virtues for their formalization. First of all, one does not take as an axiom the extremely powerful assumption that the metric has its usual pseudo-Riemannian quadratic form. Instead one assumes a group of axioms each of which is, individually, a much weaker assumption. Of course together they are enough to guarantee the existence of the usual metric form. While historically our assumption about the metric form for spacetime followed from the Pythagorean results about space established in terms of a metric usually thought of as defined by "rigid rods," these authors argue that the *generation* of this form from a number of far weaker assumptions, each testable locally in a very simple way, is a major advance.

Second, they believe their formalization is ideally suited to considering what theories might be plausible candidates as *generalizations* of general relativity. For by putting on the full metric structure a piece at a time, one can now easily generate theories more general than general relativity by asking what kind of spacetime we get as the axioms are progressively rejected, presumably starting from the last imposed and working backwards.

Now let us look at the EPS formalization from a philosophical point of view. Fortunately, we can be fairly brief, for we can avail ourselves of much of the apparatus introduced in our earlier discussion of Robb's formalization of Minkowski spacetime. As usual there will be two basic areas of philosophical interest — the EPS choice of *primitives* and the philosophical arguments for and against this choice, and the EPS choice of *definitions* and the arguments for and against them.

It is clear that Robb chose the single primitive 'after' because he felt that facts about the after relationship between events were in principle immune to claims of conventionality. EPS makes no such claim. There it is only suggested that one is deriving the full structure of spacetime from "some qualitative (incidence and differential-topological) properties of the phenomena of light propagation and free fall that are strongly suggested by experience."[29] But suppose someone were to make the claim that the construction of spacetime in the manner of EPS is "convention free." Light rays play the part of boundaries of causally connectible sets of events. And we have already given some discussion to the claim that causal connectibility is "totally nonconventional." I shall continue that discussion later.

When we move on to the role particles play in the EPS formalism, we see that any hope of totally undercutting "conventionality" theses about the metric is unsupported by this formalization. The EPS formalism assumes that we know what a free particle is. And I suppose we do, given our vast array of background theory. But to support a "nonconventionality" thesis we would need reason to believe that we could determine when a particle was free totally independently of our theoretical assumptions. But how could we do this? Certainly not by seeing that it followed timelike geodesics! And if a particle is claimed to be free, we can always "conventionally" deny this by postulating "universal forces." After all, is a particle free or not when it is gravitationally attracted by

29. Ehlers, Pirani, and Schild, "The Geometry of Free Fall," 65.

another particle? If we say it is, and use the EPS construction, we shall get one spacetime. But what if we say it is not? Won't we get a whole "conventionally alternative" spacetime by using the EPS construction?

Again, even assuming that the particle is "free from forces," we still have difficulties. First of all, we need infinitesimal particles; but that is a standard idealization. Now the choice of free particles to pick out the timelike geodesics in the EPS formalism follows because, according to general relativity, free particles travel timelike geodesics. But not all free particles do—only those that are spherically symmetric and spin-free. For if the particle has a gravitational multipole structure, it will not generally travel the timelike geodesics, even if no forces act on it.[30] And how, without already *knowing* the spacetime structure, are we to determine which particles are gravitational monopoles and which are not? There is no need here, I trust, to rehearse all the possible meanings of "conventinal" and "conventionality," nor to look at our options once again. Enough has been said to indicate that, whatever its virtues —and they are real indeed—resorting to the EPS method to determine the metric, or to any other method that utilizes free particles as well as causal connectibility, will not release us from any burden of conventionality imposed by the use of transported rods or clocks.

There may be other virtues, however, in choosing light rays and free particles instead of rods and clocks as our primitive instruments. Let us discuss one aspect of this claim here. What about the EPS definitions? Concerning the rods and clocks or chronometric formalisms, EPS argues like this: "If the g-function is defined by means of the chronometric hypothesis, it seems not at all compelling—if we disregard our knowledge of the full theory and try and construct it from scratch—that these chronometric coefficients should determine the behavior of freely falling particles and light rays, too. Thus the geodesic hypotheses, which are introduced as additional axioms in the chronometric approach, are hardly intelligible; they fall from heaven."[31] But of course the authors are

30. See V. Fock, *The Theory of Space, Time and Gravitation* (New York: Pergamon, 1959), esp. 371–374. See also B. Tulczyjew and W. Tulczyjew, "On Multipole Formalisms in General Relativity," in *Recent Developments in General Relativity.* See also Grünbaum, "Geometrodynamics," 789–791.
31. Ehlers, Pirani, and Schild, "The Geometry of Free Fall," 64.

aware that on their construction one must somehow stipulate the connection between intervals as they define them, and the behavior of rods and clocks.

The authors say that, first, one can define a "clock" by means of light rays and particles alone. Thus: "The chronometric axiom then appears either as redundant or, if the term 'clock' is interpreted as 'atomic clock,' as a link between macroscopic gravitation theory and atomic physics: it claims the equality of gravitational and atomic time. It may be better to test this equality experimentally or to derive it eventually from a theory that embraces both gravitational and atomic phenomena, than to postulate it as an axiom."

But we can test experimentally the correlation between atomic clocks and clocks constructed with light rays and particles, whether we choose the chronometric or the EPS formalization. And if our eventual overall theory allows us to derive their synchronization in one direction, it allows us to derive it in the other. Why then is the EPS method better than the chronometric?

Perhaps the authors reason as follows: if the nice agreement between atomic and light ray-particle clocks holds up forever, then it will not really matter too much which formalization we choose. But we must look forward to what we would do if we ever discovered the synchronization to break down. The behavior of light rays and free particles in spacetime is an integral part of our theory of spacetime. Their behavior follows from the most fundamental principles of our spacetime theory. But the explanation of behavior of atomic clocks depends upon theories, like quantum theory, which are not—at least not yet—so intimately related to our theory of spacetime. Should the synchronization break down, wouldn't we be more likely to look for an explanation in some aspect of the quantum behavior of atoms, than we would be to drop our fundamental postulates of spacetime theory? Wouldn't we, in other words, retain the belief that free particles and light rays travel timelike and null geodesics respectively, and the view that time intervals as measured by them are correctly measured intervals of the spacetime, blaming the discrepant results obtained using the atomic clock on some peculiarity of its makeup?

I don't know just what we would do. We might gain some insight by looking at what did happen when a generalization of general relativity was attempted but was rejected on empirical grounds.

This was Weyl's affine but non-Riemannian spacetime. This case is discussed by Ehlers, Pirani, and Schild themselves; they consider the plausibility of the additional axiom needed to go from affine to Riemannian spacetime. The authors invoke Einstein's criticism of Weyl's non-Riemannian theory.[32] The trouble with this theory was that it proposed that a vector could change length upon being transported around a closed curve. *If we assume that atomic clocks measure time intervals,* atoms should radiate at different wavelengths depending upon their histories; but they do not. This fact led physics to reject Weyl's theory. No one to my knowledge tried to save Weyl by arguing that the conformity of atomic clocks to each other was due to the fact that in traversing the closed paths they ceased measuring time properly in just such a way as to be in agreement with each other despite their journey. Nor did anyone propose "checking up" on the atomic clocks by using the pure geodesic method.

I am suggesting that, although our explanation for the periodicity of atoms may be more complex in our ultimate theory than would be our explanation for the periodicity of a particle-light ray clock, what we *mean* by 'time interval' may be more closely associated with such atomic clocks than with "gravitational" clocks. The same argument applies regarding spatial separations and rigid rods as opposed to a "gravitational" measure of spatial interval using only light rays and particles.

Suppose we discovered in the future that rods we took to be rigid, and clocks of the atomic sort we took to be ideal, no longer measured intervals as defined by the particle and light-ray methods of EPS, Marzke, and Kundt and Hoffman. What would we say? EPS et al. would have us say that we had discovered that "rigid rods" and "ideal clocks" are not very good indicators of spatial and temporal separations. From the alternative point of view, we would say that while we thought light rays and free particles traveled null and timelike geodesics respectively, they do not.

Let us consider the second position first. After all, it might be argued, just look at the history of general relativity. In Newtonian theory it never occurred to anyone, except retrospectively, to say that particles acted upon by gravity traveled geodesics in spacetime. They were "forced," and hence deviated from straight-line (geo-

---

32. Ibid., 82.

desic) motion. Now it is true that Einstein saw that one could "geometrize" gravitation because of the principle of equivalence. But wasn't his real reason for going to curved spacetime the plausibility arguments, from the red shift owing to gravity, etc., that gravity must affect our interval measurements as normally carried out by rods and clocks? If we see that spacetime as measured with rods and clocks does not coincide with spacetime as measured by the geodesic hypothesis for light-ray and particle motion, why not just drop the geodesic hypothesis? Wouldn't that be either the more plausible scientific decision—if you think there is a real decision to be made here—or, in any case, the better way of talking, if you think that all that is involved is descriptive simplicity?

But I think the reply would go like this: in our best available current theory—general relativity, the theory we are trying to formalize—the geodesic motion of light rays and free particles is a fundamental result of the theory. We cannot even separate out the field equations from the equations of motion of test objects in the field as we could in the Newtonian theory, for the very equations of motion follow from the field equations. On the other hand, the behavior of rods and clocks is a complex matter whose explanation requires not just our gravitational-geometric theory, but a quantum theory of matter as well. Now "every theory should determine its own interpretation," in the sense that the "definitions" connecting theoretical structures to observable elements in formalizations of the theory should be fundamental propositions that follow in the unformalized theory from that theory alone. On this ground, the geodesic "definition" of the metric is preferable. Admittedly, one might want to say that in adopting these definitions we have "changed the meaning of the metric terms." For, whereas the previous "fixed" propositions, invariant under theoretical change, were those connecting rods and clocks to intervals, we are now taking the "fixed" propositions to be those connecting the geometry to motions of light rays and free particles.

But we may have good reason for making this "change of meaning," if, indeed, that is what it is. We saw earlier that we could criticize Robb's definitions of the metrical quantities by seeing how the propositions connecting causal connectibility to congruence held up when we moved from Minkowski spacetime to the spacetime of general relativity. We saw that the Robbian connections were anything but "fixed point" propositions in this theoretical change. Now in critically examining a proposal to take some propositions as "definitional" in a formalization of general relativity, we have no accepted "newer, more general" theory to use as a standard. But perhaps choosing a formalization is a *proposal* about which broader theories to consider. In other words, to adopt the causal-inertial, or geodesic, definitions of the metrical quantities is to suggest that the future theories we should be considering as plausible candidates for replacing general relativity, need we do so, would be ones in which the association of light rays and free particles with geodesics is retained, whereas the association of rods and clocks with the spacetime intervals is allowed to differ from the association postulated in general relativity.

We might say, then, that whereas in going from Newtonian theory to general relativity we kept the association of rods and clocks with intervals constant, and changed our views about the association of light rays and "free" (i.e., acted upon only by gravity) particles with geodesics; in our new theoretical shifts we are likely to keep the geodesic motion of light rays and free particles intact and allow changes in the propositions which associate rods and clocks with intervals.

I think these arguments are of great interest. But I also think that they are not overwhelmingly persuasive. For one thing, the allegation that the geodesic motion of light rays and free particles "follows from general relativity alone," but that the behavior of rods and clocks requires quantum theory for its explanation, is a little misleading. To be sure, if we want to account for the fact that a particular atomic clock, say, "ticks" with a particular frequency, or the fact that a particular ensemble of atoms in a crystalline array has a particular length, we must invoke our complex quantum theory of matter. But what we need for the association of clocks and rods with spacetime intervals is not this full theory, but only a few fundamental assumptions about how clock rates, for example, *vary* with the varying gravitational potential. And the rationale for these assumptions does not use the complex quantum nature of these material entities at all. After all, the famous Einsteinian rationale for assuming that the gravitational field will have metric effects works by demonstrating that on the basis of very general considerations drawn from the equivalence principle, ideal clocks must respond, in the manner of the gravitational red shift, to changes in gravitational potential. These arguments can be classical in nature, and certainly do not depend on any detailed theory of matter we have in mind. The same holds for the usual arguments given to rationalize the belief that spatial measuring devices must also respond to the gravitational potential.[33] Now these arguments rest upon very fundamental assumptions of general relativity, perhaps even more fundamental than the field equations. So even if a theory should "provide its own interpretation," is it clear that this in any way motivates a preference for the geodesic definitions of the metric over the rod-and-clock or chronometric definitions?

Of course it remains true that if we go beyond general relativity to some newer theories, say by going to an affine but non-Riemannian spacetime, we shall still be able to retain the geodesic hypotheses intact, but shall no longer have any clear place for the chronometric. But since we do not know where we *shall* go from general relativity, is it clear that the EPS type formalisms will provide in their "definitions" the real "fixed point" propositions in our future scientific changes, and that the chronometric formalizations will not? The most we can say is that EPS formalizations are one *proposal* for a formalization best suited for the future evolution of science. If we adopt them we are implicitly assuming that in any future change we shall take the association of light-ray and free-particle motion with the geodesic structure of spacetime to remain correct; or, if we think the geodesic hypotheses are only "matters of convention," that we shall retain these conventional choices or retain this "manner of speaking." But it really is not clear, and cannot be until we have our future science, that we shall not instead hold fast to the propositions associating rods and clocks with intervals as the invariant "truths" or "conventions" or "ways of speaking."

Whatever one thinks of the EPS, as opposed to the chronometric, method of formalizing general relativity, it is still useful to contrast the arguments given above in favor of the EPS formaliza-

---

33. A very good presentation of these arguments is given by Schild himself. See A. Schild, "Gravitational Theories of the Whitehead Type," *Proceedings of the School of Physics "Enrico Fermi," Course XX, Evidence for Gravitational Theories* (New York: Academic Press, 1962), 69–115; the relevant pages are 110–115. On the general problem of the alleged superiority of one method of formalizing general relativity over another, see the remarks of J. Stachel in his review of the O'Raifeartaigh volume in which the Ehlers, Pirani, and Schild article appears, *Science* 180 (1973): 292–293.

tion with some arguments we gave earlier in favor of Robb's "definitions" of the metric. For the rationales differ in a very important way. The primary argument in favor of Robb's "definition" of the metric—and this was Robb's argument—is that it was made in completely "nonconventional" terms, using, as it did, only the allegedly totally nonconventional notion of causal connectibility as a primitive. Now, as we have seen, the causal-inertial definition of the metric throws away this advantage altogether. The metric so defined is at least as "conventional" as a metric established using transported rods and clocks.

Rather, the argument for the causal-inertial method is a "scientific" one. This definition is more "natural" and better suited for anticipated future theorizing than is the chronometric method. Whether or not we believe there are good "scientific" reasons for preferring a causal-inertial formalization of general relativity to a rods-and-clocks or to a chronometric formalization, we should clearly realize that the grounds for preference being alleged here are of a very different nature from the "philosophical" grounds offered by Robb and by others in favor of his causal theory of the metric.

Causal Theories of Topology

I have frequently promised the reader that I would return to some crucial questions about the relationship of causal connectibility and topology. If, as we have seen, a causal theory of the metric is impossible in general relativity, and correct only in a very weak sense given in the case of Minkowski spacetime, is it still possible to hold to a causal theory of the topology of spacetime in any interesting sense? And if, as we have seen, there is no escaping accusations of conventionality with regard to the metric of spacetime, is there any interesting sense in which we can at least claim that the topology of spacetime is in no way a matter of convention?

The Robbian association of causally defined features with metric features breaks down once we leave Minkowski spacetime. But what about the coextensiveness, in Minkowski spacetime, of the causally defined Alexandroff topology with the manifold topology? Does that association hold up in all spacetimes compatible with general relativity? If so, does this support the plausibility of a causal theory of the topology of spacetime?

It is now known that the Alexandroff topology will not, in fact, coincide with the usual manifold topology in all those spacetimes that satisfy the usual requirements imposed on spacetimes by general relativity. One can be sure that the two "topologies" will coincide only if the *strong causality condition* holds in the spacetime. The strong causality condition holds in a spacetime if for every point p of the spacetime and for every neighborhood of p, there is a neighborhood of p contained in the given neighborhood which no nonspacelike curve intersects more than once. Strong causality holds if the spacetime has no "almost closed" causal (lightlike or timelike) curves. If strong causality is violated, one can construct spacetimes in which the Alexandroff and the manifold topologies will not be identical.[34]

One interesting question is: should we, as a matter of principle, exclude from the realm of reasonable physical possibilities spacetimes with almost closed causal curves; or should we, rather, admit them as real physical possibilities? For our purposes the following question is more important: do causally pathological spacetimes that have almost closed causal curves block the possibility of a philosophically satisfactory causal theory of topology?

An affirmative answer to the second question is argued in this way: if a causal theory of topology is correct, we should be able to determine the topology of the spacetime using causal connectibility alone. But if we do this in the plausible way by taking the causally defined topology to be the Alexandroff topology, then if there are almost closed causal curves in the spacetime, the topology we attribute to it on the causal basis will not be the usual manifold topology. And isn't this latter what we wished to causally define?

As David Malament has pointed out to me, the existence of closed causal curves, or even of any two distinct events x and y that cause each other, not only forces a discrepancy between the Alexandroff and the manifold topology, but blocks any plausible attempt at "causally defining" the manifold topology. For if the manifold topology is defined by causal relationships it must be preserved under any causal automorphism of the spacetime onto itself; i.e., all such causal automorphisms must be homeomorphisms with respect to the manifold topology. But consider the mapping which takes x into y and y into x and leaves all other points the same. By the transitivity of causation it will be a causal automorphism; but it will certainly not be continuous in the manifold topology. The hopelessness of a program of causal definability in general is made even more manifest when we reflect on the fact that there are spacetimes in which every event is causally connected to every other event and to itself.

On the other hand, it is interesting to note that we can, by causal observation alone, determine whether or not the Alexandroff topology does in fact coincide with the manifold topology.

The causal theory of the topology seems to be foundering on the shoals of spacetimes with almost closed causal curves, just as the causal theory of the metric foundered on the rocks of spacetimes that were not Minkowskian. Now in our critique of the causal theory of the metric in Minkowski spacetimes we had to look at non-Minkowskian spacetimes. But since spacetimes with almost closed causal curves are themselves compatible with general relativity, we can attack the causal theory of topology within general relativity by using its own spacetime models.

In discussing the causal theory of the metric in Minkowski spacetime we showed that even if spacetime were believed by us to be Minkowskian, we would be skeptical of any causal theory of the metric that went beyond the mere de facto coextensionality of some causally definable notions and the metric notions and asserted some lawlike, analytic, or necessary connection between the notions. A similar line of reasoning here would lead us to assert, I think, that even if spacetime had no almost closed causal curves, the coincidence of the causal Alexandroff topology with *the* topology of the spacetime would just be a "matter of fact," and not a lawlike, analytic, or necessary truth.

We can see some of these points "in action" if we look at Woodhouse's interesting program for formalizing the causal and differentiable structure of spacetime.[35] He wishes to develop the structure of spacetime a bit at a time on the basis of axioms that have "simple and intuitively obvious physical interpretations." Further, he wishes to put the topology and differential structure on first since this will provide us with a method for characterizing singularities in the spacetime, the notion of singularity being essentially nonmetric, without presupposing the whole metric structure.

Woodhouse wants to consider only spacetimes that have no almost closed causal curves. How does he do this? He introduces as

35. Woodhouse, "Differentiable and Causal Structures," 496–497.

primitive the set of events and *particles*. The particles, once again corresponding intuitively to possible paths of free particles, are assumed to be one-dimensional subsets of M, the set of events that have a continuous structure and are homeomorphic to the real line. So closed timelike lines are excluded ab initio. Using the topology of the particles, built into the primitives, it is possible to introduce an axiom equivalent to assuming the spacetime to be free of almost closed causal curves. Having done this, one can introduce the Alexandroff topology and take it as the topology of the spacetime. The topology so defined induces the already assumed topology on the particles, and as we shall see later, has other "virtues" as well. The important thing to note here is the necessity to first assume some topological property as a primitive in one's formalization before a "causal" introduction of the full topology even becomes possible.

Even if the possibility of spacetimes with almost closed causal curves had not occurred to us, I think we could still have some reason to be cautious in accepting a causal theory of topology. In considering earlier definitions of the metric we asked the question whether the definitions in a particular formalization could be considered right or wrong. If they could — if there were some criterion of rightness or wrongness for some "definitions" — we saw that one might hesitate to call these "definitions" really definitions. For it seemed that we had some check on the correctness of the correlations asserted by the "definitions," and this suggested either that these "definitions" were *factual* propositions, or at least that we had some other notion of the *meaning* of the defined expression in mind. Thus if we can say that X is or is not *correctly* defined by Y, it would seem that we already had some notion of what X meant.

We also saw that in looking at causal "definitions" of the topology even in the case of Minkowski spacetime, more than one such causally definable "topology" would be constructed. It then became a question of which "topology" was *the* topology of the spacetime. Once again, what criteria could we use to judge this if our only notion of a topology was that which the causal "definition" introduced?

In the case of causal topologies for Minkowski spacetime, for example, Zeeman was careful to indicate that his causal topology, although it differed from the usual Alexandroff — and hence from the usual manifold — topology, did, like the more standard topology, induce the usual three-dimensional Euclidean topology on spacelike hypersurfaces and the *usual one-dimensonal real line topology on time axes.* If a causal topology did not do that, would we even consider it as the "right" topology for Minkowski space-time?

Woodhouse also takes as a check on the "correctness" of the Alexandroff topology the fact that it induces back on the particles the initial topology presupposed for them, that of the real line. Further, he shows this: if a sequence of events, x_n, converges to some given event, x, in the Alexandroff topology, and if p is a particle through x, then the time interval (on p) needed to travel from p to x_n and then back to p goes to zero as n goes to infinity. So, he says, there is "an obvious physical interpretation of the convergence."

Perhaps the following is suggested by these points: we may not really have any a priori idea of what the full topology of spacetime should be. Witness the difficulty of trying to decide whether the correct topology for Minkowski spacetime should be the Alexandroff or the Zeeman. But we do believe we have a *primitive* idea of what a continuous particle motion, and hence a continuous time-like line, is. Whatever topology we impose, causally or otherwise, on the spacetime, we can judge it to be correct or incorrect depending on whether or not it induces on the particle paths the continuous topology we had in mind in the first place.

This suggests that if anything is a "nonconventional" fact about the spacetime, it is whether a given timelike connected set of events is continuous or not. So at least *that* topological notion should be a primitive in any formalization. And, of course, once we have this, the notion of a genidentical particle can be introduced by *defining* a genidentical particle to be a continuous timelike world-line. Thus while a causal theory of topology seems ruled out, at least if we are going to allow ourselves to countenance spacetimes with almost closed causal curves as *intelligible* (we need not believe them physically possible for the argument against the causal definition of topology to apply as an argument against the causal "definitions" being analytic or necessary, at least), perhaps some version of a topological theory of causal connectedness is in order.

What I am suggesting is that in any full formalization of a space-time theory the notion of a continuous particle path will be assumed as a primitive. Like facts about the incidence or nonincidence of events, facts about the continuity of particle paths will be, in Reichenbach's terminology, "elementary facts." Insofar as we take aspects of our theory to be conventional, in any of the senses of that much abused term, incidence and continuity of particles will be "nonconventional" consequences of a theory.

I do not know what makes a term primitive. I have already expressed my doubts that "direct apprehensibility" will be the criterion, because, again following Reichenbach, this might force us to take our primitives as "items of subjective awareness" and lock us forever behind the veil of perception. All I am suggesting here is that if some notions are primitive, at least one topological notion, that of continuity of a possible particle path, should be one of them.[36]

VIII ONTOLOGICAL QUESTIONS

The questions discussed in this chapter have been primarily epistemological and semantic: how do we know what spacetime is like? To what extent is this a matter of convention? How do the terms of the theory of spacetime get their meanings, and what are these meanings? To what extent can alleged definitons of terms be criticized and judged right or wrong? Let me make a few brief ontological remarks before I conclude.

Suppose we accepted some definition in a spacetime formaliza-

36. David Malament has shown that we need assume only past- and future-distinguishingness in order to be sure that the manifold topology is at least implicitly causally definable. See his Ph.D. dissertation for Rockefeller University, "Some Problems," and "The Class of Continuous Timelike Curves Determines the Topology of Spacetime," *Journal of Mathematical Physics* 18 (1977): 1399–1404. Of course if the spacetime is not strongly causal the Alexandroff topology and the manifold topology will still not coincide. (For relevant definitions, see Hawking and Ellis, *Large Scale Structure,* 192.) Malament has also shown that the class of continuous causal curves "defines" the topology in the sense that given two spacetimes equipped with the usual manifold topologies, if there is a one-to-one function between them taking continuous causal curves into continuous causal curves, this function will also be a homeomorphism. This result holds generally in spacetimes. Again we see that while causal connectibility will not, in general, "define" the topology in even the weakest possible sense, the notion of continuity along causal paths will come closer to doing the job. But even in this case much caution is called for if one is to decide in just what senses even the class of continuous causal curves does or does not "define" the topology. For more on these matters, see L. Sklar, "What Might Be Right about the Causal Theory of Time," chap. 9 in this volume.

tion, taking it that to be an "X" meant to be a "Y." What would this say about the reality of "X's" of their "ontological reducibility" to "Y's"? Suppose, for example, we define 'timelike geodesic' as 'path of a free particle.' What does this say about the reality of timelike geodesics, or about the reducibility *ontologically* of spacetime entities to material particles? Notice that in most formalizations in which such a definition would appear, it is usually taken for granted that the world is full of particles; essentially we act as though every timelike geodesic was actually traversed by some *actual* free particle.

But, of course, in the real world we do not believe this. What we really believe, if we accept this definition, is that every timelike geodesic is the path of some *possible* free particle. So instead of saying "X is a timelike geodesic if it is the set of events occupied by the history of some free particle," we should really say that "X is a timelike geodesic if it is the set of spacetime locations that *could* be the locations of the events constituting the history of some free particle."

Unless we want a metaphysics replete with "permanent possibilities of location," ungrounded on any actual locations that might be occupied by events but just are not, the natural way to understand this is in a substantivalist, not a relationist, ontology of spacetime. Just because we believe that the meaning of 'Y' is definable in terms of 'X,' where Y's are spacetime entities and X's material events, and just because we believe that our only epistemic access to Y's is through X's, we are not committing ourselves to relationism and eschewing a realistic and substantivalistic theory of the metaphysics of spacetime.

IX FACTS, CONVENTIONS, AND ASSUMPTIONS IN THEORIZING AND FORMALIZING

I suppose all but the most adamant apriorist would admit that we must rely upon *observational facts* in deciding, on a scientific basis, what the spacetime of the world is like. Insofar as we at least *seem* to theorize beyond the realm of pure observational testability, we shall be open to the suggestion that we must make conventional assumptions as well, although, of course, disputes will arise over whether or not these conventions are merely "trivial semantic conventions" or whether they are "real choices unconstrained by observational data"; and over whether or not we have grounds beyond the observational data for making the choices in a rational rather than merely conventional way.

Many people would now admit, I believe, that more is needed as well. For, they would agree, before our decision-making apparatus can even begin to make use of the factual and conventional material we feed it, we must add a number of *theoretical assumptions* about just what theoretical options we consider open as real possibilities for being a correct theory of the world.

My aim in this paper has been to argue that the choice of formalization of a theory is replete with scientific and philosophical consequences. For this reason, we must utilize our full resources in determining facts by observing, making conventional choices, and acting within a framework of assumed theoretical options in order to choose a formalization that is rational on scientific and philosophical grounds.

Our characterization of terms as primitive or defined, and our characterization of consequences as definitional or postulational, are decisions that implicitly reveal our beliefs about the limits of observational testability of theories and about the ability of theories to outrun these limits in their content; our ideas about the meanings of theoretical terms, how they are fixed, and how they change; and our views about the place of the theory formalized both in the historical context of the theories which preceded it and from which it evolved and in the assumed future science which, we anticipate, will perhaps evolve from it under pressure of new observation and new theorizing.

Given that so much hinges upon the formalization we do choose, we should not be surprised to find that this choice is no trivial matter, but one that requires the full utilization of our best available scientific and philosophical methodology. If formalizing is to be more than mere "logicifying," it *is* theorizing, and demands, if it is to be done adequately, the full resources needed for theorizing in general.

4. Do Unborn Hypotheses Have Rights?

This piece examines some variants on and responses to a skeptical challenge to our claim to have justified belief in science. The form of skepticism entertained is not that which rests upon the fact that we have never examined all possible evidence, nor on the general skepticism sometimes propounded which rests upon the fact that generalizations are never logically implied by the particular evidence for them no matter how extensive that evidence might be.

Rather, I am concerned with the skeptical doubts which might arise when one realizes that all theory choice is selection of a best alternative from hypotheses considered, and that the finiteness of our imaginative procedures always leaves an infinity of alternative conceivable hypotheses never even thought of by us.

Various grounds for rejecting the alleged skeptical consequences of this view are examined, including the general objection to invidious comparison of real procedures with impossible ideals, and various a priori and empirical arguments to the effect that the theories we have considered are, at least in probability, generally better candidates for belief than those we have not yet imagined. The differential reaction to this argument engendered by reflection upon, on the one hand, the possibility of unimagined alternatives to our most fundamental world-hypotheses and, on the other, to the possibility of alternatives to our more modest fragments of limited generalization in specific sciences is noted and explained.

Some examples from the history of spacetime theories are used to motivate the consideration of the possibility of hypotheses not yet imagined, but better in their intrinsic nature than any of those we have thought of, as being a problem worth our serious attention, and not just a philosopher's whimsical logical possibility.

I

OUR intuitive satisfaction with philosophical models of rational belief in science is, of course, highly influenced by our
experience of the historical development of science as practiced.
While one can go overboard in this direction, emphasizing conformity to a history infected with "accident" and neglecting critical
examination of the usual standards of rationality on their own
merits, surely it is reasonable to let one's views of how science ought
to be influenced by how best scientific practice has historically
proceeded.

Clearly, the long existence of geometry as the single clearly sys-
tematized and secure science was largely responsible for the persuasiveness of models of science which posited that (at least ideally) all
beliefs ought to be founded upon indubitable self-evidence. The
lengthy sway of Newtonian mechanics as a (possibly) all encom-
passing and final theory founded upon inference from the data of
the senses surely was instrumental in the long domination of philosophy of science by inductivist models. And the scientific revolutions of the twentieth century, especially those aspects of the revo-
lution which showed us that the wealth of data previously taken to
support Newtonian theory was, when taken in conjunction with
the new data incompatible with the older theory, equally supportive of novel theories incompatible with the Newtonian, was
straightforwardly influential on a range of philosophies of science
ranging from Popper's skepticism of the notion of inductive support through those who take the quasi-Kantian line that our very
understanding of what the data are is an imposition on the world by
us of our theoretical preconceptions.

The question I want to examine here is this: Quantum mechanics and relativity show us that the data which seemed to support very well one theory can be seen, in the light of scientific
change, to support as well quite different theories, theories incompatible with the older theory in deep theoretical and conceptual
ways. Now consider the attempt to decide, at any moment, whether
or not a given set of data supports a hypothesis. Having in mind the
potential existence of a vast array of theories, genuine alternatives
to the hypotheses we have brought to mind, but alternatives which

we haven't yet even imagined, should we not wonder if among that vast array of alternatives there might not be some which, were we to know what they were like, we would consider even better supported by the present data than any of the alternatives we have brought to awareness?

This is, of course, a kind of skepticism about induction, but it is worthwhile to distinguish it from skepticism founded upon distinct, although closely related, grounds. We can look upon Humean skepticism, perhaps, as skepticism founded upon the fact that all the possible data are never in. I ought not to believe all crows are black on the basis of the observed sample of crows, for nothing in the nature of the sample assures me that the very next crow won't break the pattern.

We could be led to a skepticism closer to the one I have in mind by Goodmanian arguments, although Goodman, of course, doesn't draw skeptical conclusions from his observations. "Grueish" hypotheses are, like the ones I have in mind, alternatives to our ordinary hypotheses all equally compatible with the empirical evidence to date. But, presumably, what we are to do as philosophers is to figure out what is wrong with these outlandish hypotheses. Whether the line one takes is Goodman's notion of the entrenchment, or some version of the thesis that our selection of projectible predicates is itself rationalized by our current best available background theory, we are, in any case, led away from skepticism by some argument designed to show us that as things stand now we do have some reason for preferring as genuine scientific alternatives members of the set of hypotheses from which we do choose, to alternatives constructed in the Goodmanian manner which can somehow be rejected as spurious candidates.

While the issues here are close to those with which I am concerned, they are not identical. What I have in mind is, rather, the skepticism engendered by reflection upon historical scientific experience; skepticism based not upon the existence of outlandish pseudohypotheses cooked up by the philosophical manipulation of predicates, but, rather, on the reasonable assumption, warranted by past experience, that there are vast numbers of perfectly respectable scientific hypotheses, hypotheses which, were we aware of them, would receive our most serious scientific consideration, but which, due to the limitations of our scientific imagination, we just haven't yet brought to mind. What we are concerned with is a

Newton dubious of the inverse square law, not because objects might obey it up to 1700 and cease to do so thereafter, but because he imagines the possibility of an array of genuine alternatives to his theory even though, of course, he can't imagine just what such alternative theories would be like.

Nor am I concerned with the problem of theoretical underdetermination in any of its forms. The alternatives I have in mind are not those variants of the original theory which a positivist would declare trivial semantic alternatives. I shall not be concerned with alternatives constructed by manipulation of the theoretical apparatus which leaves observational consequences invariant, nor with those quaint alleged alternatives which one gets by switching from talk of objects to talk of time-slices, from things to modes of space-time points, and the like. General relativity is not, in any way, an "empirically equivalent" variant of the Newtonian theory of gravitation. But shouldn't a prudent Newton have realized that such a theory could exist, even if he couldn't say what it was? And might not some such unimagined theory be more plausible, even relative to present data, than Newton's own? And shouldn't we now admit the existence, Platonically, of innumerable alternatives to our best present theories, alternatives all of which would save the current data equally well, but none of which are equivalent, vis-à-vis all possible empirical experience, to the currently accepted hypotheses? Shouldn't we realize that these alternatives exist, even if we can't say what they are? And shouldn't we accept the fact that among these unimagined alternatives some may very well be more plausible than our own theories relative to present observational facts?

The depth of the problem we face here can be emphasized by just a brief consideration of a number of models of theory choice we have been offered by methodologists. In each case we see that we can make sense of the adoption of one hypothesis only by viewing the decision process as the selection of the preferred hypothesis from an antecedently given set of possible alternatives. In each case we must wonder if we are left with any coherent theory at all when we face up to the fact that the set of hypotheses we have yet brought to mind constitutes only an infinitesimal finite fragment of the full range of alternatives.

Consider, for example, the attempt to resolve the familiar problems encountered in trying to define a notion of qualitative confirmation which proceeds by arguing that an instance is confirming of one hypothesis only relative to an alternative choice or set of such alternative choices. If the set of alternatives we should really have in mind is the indeterminate class of all possible hypotheses, including ones we have not yet thought up, can we understand what it is for an observational instance to confirm a given hypothesis at all?

Consider Bayesian strategies for confirmation theory. Here we must distribute a priori probabilities over all the alternative hypotheses to be considered. If there is only a finite set of hypotheses we have in mind, this is easy to do, even if it isn't easy to find any source beyond subjective whim for rationalizing any particular chosen distribution. Even when the set of alternatives is infinite, if it can be characterized by some orderly parameterization we have the means (say by moving from point to interval estimation) of plausibly assigning a priori probabilities and then grinding through conditionalization with respect to the evidence in the usual way. But if we must keep in mind the infinite and *indeterminate* class of all possible hypotheses, known and unknown, how can we even begin to assign a priori probabilities to those few hypotheses (or parameterized sets of them) we do have in mind (unless, perhaps, to give them all a priori probability zero on the basis of their very small place in the space of all possible hypotheses)?

Consider inference to the best explanation. Should we adopt that hypothesis relative to which the evidence has the highest likelihood of all the likelihoods generated by the hypotheses we have in mind as alternatives? Rather, shouldn't we realize that in the vast sea of alternatives we have not yet considered it is all too probable that there is some, as yet inconceived, hypothesis relative to which the evidential warrant is even better explained than it is by our current best candidate? On this basis shouldn't we agree that being the best of an arbitrarily selected and narrow class simply isn't being good enough to be believed, and once again skeptically withold our judgment?

Finally, we ought to consider those attempts at reconstructing scientific inference which rely less upon formal models of hypothesis choice and more upon models of choice allegedly founded upon abstraction from historical scientific practice. Familiarity with this literature once again shows us a universal predilection for the competitive model: belief is to be credited to that hypothesis which does best in competition for survival with its rivals, be they older, previously accepted, hypotheses or novel alternatives recently contrived. But what credibility can accrue to the victor in a battle for survival which, by historical accident and paucity of imagination, simply keeps nearly all of the competitors out of the arena?

II

An initial response to the alleged skeptical consequences of the existence of inconceived hypotheses would be to affirm that since we cannot possibly deal with the unavailable, its existence or potential existence cannot be relevant to questions of justification. For example it might be argued that to be skeptical of induction because all of the data on which we might make a judgment are not yet in is to misconstrue what justification in the inductive context means. The whole point of induction is to allow us to draw inferences on less than an exhaustive observation of all the facts in the world, and so, to declare an inference unjustified simply because it proceeds on the basis of a small sample of observational facts is to misconstrue what would count as a justified inference in the inductive context.

Similarly, the argument might go, by its very nature the process of hypothesis selection in science requires us to make our decision on the basis of the consideration of only that limited selection of hypotheses which have come to mind at a given time. It is as impossible to consider as potential candidates for belief all possible hypotheses as it is to have in view all possible observational facts about the world. Accusations of "mere contingency" or of its being "accidental" that we have considered those hypotheses we have brought to mind and not the others are not more supportive of skepticism here than would be the claim that it is accidental that we have the observational data we do have and not some other sample in the inductive case. To be justified in accepting a hypothesis, it will be argued, is to have selected the best among the available candidates. The invocation of other possible hypotheses not yet brought to mind as a grounds for skepticism is just a misconstrual of what counts as justification in the context in question. It is no argument that one isn't justified in doing what one does on the basis one has by pointing out that the basis for inference is weaker than another which is, in fact, impossible to obtain.

Up to a point I think that this argument has merit. Whenever a procedure in which we have preanalytic confidence is criticized by invidious comparison with some ideal which is, by its very nature, unobtainable, we are justified in invoking as our first defense of the procedure in question the argument that the only justification for giving up a procedure, imperfect as it may be, is the move to a better one. It is never justified, it will be claimed, to give up the best procedure one has simply on the basis that it doesn't meet standards which are in principle unfulfillable in any case.

But, of course, the skeptic can reply that just as the best hypothesis may not be good enough, leading us to withhold judgment (and this not simply as a skeptical general withholding of judgment, but as the right thing to do, at least temporarily, in actual cases — even under very unskeptical decision-making schemes), so the best methodology may simply not be good enough. If this leaves us without any rational grounds for decision-making, says the skeptic, well, that is just too bad. Of course we may continue to decide with the vulgar while chastely withholding judgment in the privacy of our learned study.

Rather than debate this general issue, though, I think it more profitable to note that there are some general moves one can make in this case, moves which if not fully resolving the issues between the skeptic and his opponent, at least throw some light on the issues between them. Let me turn to these.

III

First let us consider a couple of related arguments which attempt to mitigate our skepticism by urging on us a subtler understanding of the *content* of our hypotheses than might have initially occurred to us.

(1) One such move is to urge on us a distinction between our fundamental theory of the world (or "cosmology") and those more limited theories characterizing narrower features of reality. Into the former class go such things as our general spacetime theory, our general theory of objects and their states (at the moment relativistic quantum field theory — such as it is), etc. Into the latter class would go theories of the chemical realization of the genetic code, of covalent chemical bonding, of superconductivity, of plate tectonics, etc.

Do we really harbor the suspicion that, in the fullness of time, we will either come to believe, aware as we will be of novel hypotheses we hadn't previously considered, that our earlier acceptance of theories was then unwarranted? Of our general, fundamental, cosmological theories I think this might be so. And if we do harbor the suspicion of this possibility, how can we be assured of the permanence of the warrant for the narrower, less fundamental theories all of which presuppose the correctness of our cosmological views?

I suggest it is just this presuppositional nature of the narrower theories, properly understood, which can immunize them from the kind of skeptical doubt which infects our belief in our grand cosmology. For, I suggest, we may understand the content of the narrower theory as being relativized to the background cosmology. So that our understanding of, say, the claim that genes are DNA molecules can be thought of as invariant (and hence its warrant invariant) under even a radical revision of our understanding of just what molecules (or material things in general) are.

Suppose Wheelerian fantasies are one day realized, and we find that DNA molecules, like all other matter, are just tightly knotted up bits of spacetime. Even wilder, suppose that our future scientists finally see the wisdom of the Leibnitzian view that all the world is a construct of spiritual monads. Would any of this change our opinion that genes are DNA molecules? Or that we were so warranted in believing in 1975? Would it change our estimate of our earlier warrant that electron sharing holds some molecules together? Or that superconductivity is the result of long-range pairing which bonds fermion electrons into the suitable constituents of a degenerate boson gas? I think not.

Of course how we then "understand" those narrower theories might change. The meaning of plate tectonic theory is, I suppose, rather different when we come to suppose that crustal plates are, like everything else, thoughts in the mind of the great world-spirit (or whatever), than is the meaning of that theory under our present cosmological scheme. But be that as it may, I still think there is plausibility to the claim that while contemplation of inconceived hypotheses might make us skeptical of our warrant for our grandest, most fundamental theories, it really doesn't influence our confidence in the warrant we hold for inductively well-grounded narrower theories. Do we really think that any future increment in scientific imaginativeness could lead us to doubt that yeasts cause fermentation, or that vision is mediated by the impact of light on the retina of the eye? If the future does do in these theories it will do so only by a dissolution of the cosmological background they presuppose. And if we understand the meaning of these narrower theories as being relativized to the background cosmology ("Genes are DNA molecules, whatever molecules are"), we block skeptical concern about the narrower theories based on the dubiousness of their cosmological backgrounds.

Nothing here is meant to prove that we ought not to be as skeptical of the warrant for the narrower theories as of the grander. Rather, it is meant to point out that insofar as there is *real* doubt, based on the contemplation of hypotheses yet unborn, it is doubt localized in general in the domain of fundamental theories. Such real doubt infects our confidence in the warrant of narrower theories only to the extent that they are not yet firmly established in the ordinary, inductive, sense. No doubt we are more confident in the theory of polymers than we are, say, in the theory of superconductivity or of plate tectonics. To that extent we remain more open to the possibility that there is some real, alternative, better explanation of the phenomena treated by the latter than we do to the possibility that some alternative to long-chain molecules will someday explain plastics better than current molecular chemistry does. But, in general, while both narrow and cosmological theories can be replete with empirical support, for the latter, but not for the former, we are genuinely assailed by doubts which originate in our awareness that there are more theories in Plato's heaven than have, as yet, been dreamt up by our theoreticians.

(2) There is another way of attempting to vitiate skepticism by reunderstanding the content of theories which is generally applicable to both narrower and cosmological theories. The degree to which it can assuage our skeptical doubts is, however, quite problematic. Here I refer to the proposal that we take hypotheses not as asserting their manifest content but only as putting forth that content as approximately true or partially true.

Perhaps this approach can be motivated by referring, once again, to the downfall of the Newtonian system. As everyone knows, wrong as Newtonian dynamics and Newtonian gravitational theory may be, the theory remains a superb approximation to the truth in a vast variety of instances. Should we not, then, in anticipation of future scientific revolutions overthrowing our present cosmological theory, take our present fundamental theory to be, like Newton's, at best a body of partial and approximate truth? And if we do this can we not protect ourselves against the attack of the skeptic? For while his allusion to unborn hypotheses may very well make us dubious of our claim to hold the final truth, even the prospect of unimagined alternatives to the ones we have considered cannot lead us to doubt that our present best cosmological theories are warranted as at least a good approximation to the truth or at least in part true.

One objection to this "way out" might be that although we can, indeed, have some assurance that our present best theories will always retain some value as approximate truths, we cannot now know, in the light of future hypotheses we might come up with much less in the light of possible new data, just where the present theory is going to fail to be reliable (or warranted as reliable) and where it will retain its approximative value. Thus, it might be argued, even weakening our attribution of content to one of mere approximate truth does not really help us in the face of skepticism of the kind we have been considering, since we can't now know, in the absence of knowledge of just what hypotheses are as yet unborn, just which parts of our current theory we are presently warranted as taking as even approximately true.

But, realistically, things aren't as bad as all that. Can't we be assured now, that whatever new hypotheses come along in the future, we will never give up our present estimation of the reliability of relativistic quantum field theory to predict the energy levels in the hydrogen atom? Couldn't Newton, ignorant of quantum mechanics and of relativity but knowledgeable of the fact that some theories he hadn't yet thought of would fit the data as well as the best theory he had come across, still be assured that whatever the future brought, the inverse square law of gravity and Newtonian dynamics would remain inviolate as a good approximate account of planetary motion?

There is a far deeper objection, however, to the way out of skepticism which tells us to take the content attributed to presently believed hypotheses as only approximate and partial. Once again the real pressure of skepticism only arises when we consider our most fundamental, cosmological theories. It can be argued (and frequently has been) that the transition from Newton to relativity (not to speak of the transition to quantum mechanics) constituted a total shift to a new world picture completely incommensurable (as the cliché goes) with the picture of the world on the older theory. From this, very familiar, point of view there is no sense in which the

older theory can be viewed as even approximately correct, miscon-
struing as it does (from the relativistic point of view) the most
fundamental facts about spacetime, and, worse yet (from the quantum mechanical point of view), being mired in the fallacious classical conception of an objective world independent of measuring
apparatus, etc. Even as an approximation if we could make sense of
such a notion, Newton's theory is a failure.

Seeing this, how could we not imagine the vast realm of hypotheses we have not yet thought up which differ as much in their ultimate picture of the world as our present physical picture differs
from the Newtonian? And yet each of these hypotheses is to be
imagined as fitting the data by which we support our present account as well as does our own best theory and better supported by it.
Imagining such a realm of not yet conceived hypotheses, how can
we be anything but skeptical of the warrant by which we hold to
belief in our present theories—even viewing them merely as approximations to the truth?

Perhaps there is an antiskeptical reply to this. It would require
first making sense of the notion of conceptual approximation,
showing us, against the familiar arguments of the incommensura-
bilists, that one can make sense of one theory approximating the
conceptual apparatus of another. Next the full reply to skepticism
would have to indicate some way in which we could obtain some
assurance that we need not now fear the existence, in the realm of as
yet inconceived hypotheses, of numbers of sufficiently plausible
hypotheses so radically different conceptually from our present
best theory that we could not speak of our present best choice as
even conceptually approximating these new alternatives. For were
such unborn hypotheses to be now believed in by us, would we not
be skeptical of the warrant for our current best theory? I have no
idea how either a notion of conceptual approximation could be
obtained, nor any idea how we could assure ourselves of the nonexistence of plausible but radical still inconceived novelties.

IV

There is an alternative method of attempting to evade skepticism
which allows us to credit to our hypotheses the full content they
seem manifestly to maintain. This option suggests, rather, that we
modify our epistemic attitude toward the "winning" hypotheses in
some direction away from the intuitive and naive notion of belief.

That is, instead of saying that we believe the most successful of contending hypotheses (withholding belief, presumably, because of our awareness of the rich body of not-yet-conceived alternatives), we say, rather, that we hold some weaker, subtler epistemic attitude toward the victorious contender.

One such approach, of course, is to move from the dichotomy of belief and disbelief toward some version of epistemic probability —subjective, logical, or otherwise. But I don't think that this will be a satisfactory move at this point. For, as we have seen earlier, it would be too easy for the skeptic to push us to the point of admitting only zero probability for the hypotheses we have thought up. There is an antiskeptical reply to this to which we will return later, however.

The approach I have in mind here, though, is to adopt a locution similar to those familiar to us from Popper. Since the totality of data is never in, and since Popper is skeptical of inductivist claims to warrant from samples, he argues that we ought not to believe even our best conjectures. Rather, we ought to speak of "tentatively holding" them or "momentarily adopting" them. Should we not, similarly, in the face of unborn hypotheses, forever withhold belief from our best current contender, instead adopting a fallibilistic attitude framed in some such jargon?

While there may be some value to this approach, in the end I think it the least useful way of confronting the skeptical challenge. The trouble is, of course, that just as "all the data" are never in, so we never exhaust in our consideration the totality of the Platonic universe of possible hypotheses. On the view under consideration, then, we ought to forever withhold belief, however partial. But if our stance is, forever, merely to adopt or maintain (or whatever) a current hypothesis, but at the same time to use it to explain the phenomena, predict the outcome of future experiments, and control the world, then what useful distinction have we made between this allegedly weakened epistemic attitude and genuine belief? Surely, moving in this direction smacks all too much of wanting to have one's skeptical cake while getting the nutritional advantages of consuming the meal, dessert and all.

V

The most interesting reply to the kind of skepticism we have been considering is that which attempts to meet it head on. Rather than evading skepticism by weakening the putative content of hypothe-

ses which are accepted from their face value, or by reducing acceptance to something less than belief or partial belief, this approach
argues that we might very well have good reasons for standing by
our choice of the best of the contenders for belief, even acknowledging the vast array of as yet unexpressed alternatives we have ignored
in the decision-making process. These reasons might be "quasi a
priori" or they might, themselves, be a sort of induction from our
empirical experience. Let us look at a few arguments of this sort.

(1) What could assure us of the reasonableness of our belief in the
hypothesis selected as best from the set of available contenders even
though we know there is an infinite plentitude of alternative hypotheses we simply haven't considered? Well, suppose we had
some reason to believe that the hypotheses we have considered are
all, in at least one crucial respect, and relative to present data of
course, superior in warrant to all those not yet thought up. But how
could we know or have reason to believe this when we don't even
know what the unborn alternatives are?

Consider the *simplicity* of hypotheses. Whatever that is, and
however we are to assign degrees of it to hypotheses, it is frequently
said that, all other things being equal, we ought to believe the
simplest of alternative hypotheses. Why we should prefer the simple hypothesis to the more complex is very hard to say. Without
some plausible apriorism it is hard to connect simplicity with any
plausible "mark of truth," and "pragmatic" rationales are notoriously difficult to sustain in the epistemic context. Be that as it may,
let us assume that it is right to believe the simpler rather than the
more complex alternative (all other things, of course, being equal).

But if simpler hypotheses are preferable to less simple, do we not
have at least some reason for dismissing the claims of skepticism
founded upon the existence of as yet inconceived hypotheses? For
do we have not, now, some fair warrant for believing that the
hypotheses not yet thought up will be less simple than those which
have come to mind?

By what right might we believe this? One reason could be the
"empirical generalization" that, as a matter of psychological fact,
scientists generally do think up simple hypotheses before the more
complex alternatives occur to them. More interestingly, if more
speculative, is a "quasi a priori" argument: Perhaps the very meaning of simplicity is given by the order of imagination. Perhaps we

simply call simpler those hypotheses we are (generally) likely to think up first. From this point of view simplicity, and hence plausibility, are granted to those hypotheses we have considered over those we have not by the very fact that the former have come to mind!

Here resort to historical cases may be informative, but the matter would have to be explored with some care. To be sure, the data which Newton relied on in support of his theories of dynamics and gravitation can be seen to be compatible also with quantum mechanics and with special and general relativity. But, surely, even were Newton to be aware of those theories he would have been justified in sticking to his original hypotheses as the simplest, hence most worthy of belief, of the alternatives then in mind relative to the data available to him.

On the other hand, there are some well-known alternatives to the Newtonian theory, unavailable to Newton, which are such that we might very well be inclined to say that had Newton been aware of them he would have been obligated, by his very own standards of rational belief in science, to have opted for these alternatives and dropped his original theory, even relative to the data which he was in possession of at the time. I have here in mind so-called neo-Newtonian spacetimes in both their flat and curved version. The former provides an alternative to the Newtonian account which allows one to retain the empirical consequences of Newtonian dynamics without postulating a spacetime structure rich enough to allow the definability of absolute velocity, a notoriously "unobservable" quantity in Newton's original theory. The latter assimilates gravity to spacetime curvature, as in general relativity, but in such a way as to reproduce the empirical consequences of Newtonian gravitational theory. Again it has conceptual advantages over the original Newtonian theory, eliminating the distinction, present in the original theory, but empirically undeterminable, between the absence of a gravitational field and the presence of one whose uniformity makes it undetectable by fooling us into thinking that free fall is inertial motion.

It is at least plausible to argue that had Newton been aware of these alternatives, considerations of simplicity (in some sense of that elusive concept) would have obligated him to prefer them as explanatory of the very data he used to back up his belief in his own theory. But here the alternatives to the original theory are those a positivist would declare mere trivial semantic variants of the older theory, for they are alternatives specifically designed to duplicate the totality of observational consequences of the original Newtonian theory.

It would be well worth a historical investigation to ask if there are any cases where the following pattern of development actually occurred: An original theory was well accepted by the scientific community to account for a range of empirical data. Later a new hypothesis was thought up which, while having some predictive consequences which have distinguished it as a genuine alternative theory from the original, was equally compatible with the original vis-à-vis the range of data which was taken to support the earlier theory. Further, the alternative was such that, even neglecting the possibility of testing it against the original theory by empirically exploring those regions where they give differing empirical predictions, a rational person would argue that had this theory been thought up at the time, it would have been more worthy of belief than the older theory at the time the original theory was adopted and in the light of the data which at that time supported the theory actually adopted.

I have not been able to think of a clear-cut example of this kind. Perhaps there are some. But there are surely few such cases. Isn't that alone somewhat persuasive in favor of the argument we have been considering: Either as a matter of empirical fact or of a priori truth, we just do think of the more believable hypotheses first. Yet the example of neo-Newtonian spacetime, although, as explained, not quite the kind of example needed to counter the claim, will make one hesitate to accept this refutation of skepticism too glibly.

A variant of this account might go as follows, bringing us once again to the problem of a priori probabilities considered earlier: In any Bayesian theory of inference we need to assign to hypotheses intrinsic a priori probabilities. Could we not have some grounds, again either empirical or a priori, for assuming that hypotheses we have thought up have higher a priori probabilities than any in the vast array not yet brought to mind? Once again this belief could be founded on a psychological (historical?) generalization that people generally do think up intrinsically plausible hypotheses first. Or it could be founded on an attempt to show that our very assignment of a priori probabilities, on many accounts a "subjective" matter anyway, is just a numerical representation of the order in which hypotheses occur to us. Even the infinitude of hypotheses not yet brought to mind need not be devastating to this argument, since we could supplement it with some view about the way in which a priori probability clumps up among a few initial alternatives leaving little to be distributed among the vast totality remaining. Once again, a defense in depth of this position would not be easy to provide, but is there not some plausibility to the claim that although there are, indeed, many, many hypotheses we haven't yet considered, surely there is sometimes good reason to think that the ones we have brought to mind grab the lion's share of intrinsic believability among them?

(2) A related but interestingly different argument against the skeptic would have us rely not upon the general likelihood of simpler or more a priori plausible hypotheses coming to mind first, but, instead, upon a higher-order inductive rationale to the view that, as things actually stand now in the present state of science, we need not be reduced by the spectre of unborn hypotheses to skeptical dismay.

The argument I have in mind goes something like this: At an early or "immature" state of science we should, to be sure, be extremely tentative in adopting any hypothesis merely because it is best among the competitors brought to consciousness. But as science enters its mature stage we need not be so diffident. On the basis of a kind of metalevel inductive reasoning we may have good reason to believe that the hypothesis on which we have fixed our belief is not only the best possible choice from among the alternatives we have considered, but is also superior to any hypothesis in the remaining body of those as yet unimagined. Indeed, we may have reason to think that of all those as yet not brought to mind, none is sufficiently viable that were it to occur to us it would result in any serious weakening of our confidence in the choice we have made.

Actually, the usual version is a little more modest than that. More often it would be claimed that while awareness of a novel hypothesis would make us lose some confidence in our present beliefs, we would still be able to retain great confidence in our currently accepted beliefs as good approximations (conceptual and otherwise) to the truth. That is, while few would argue that such higher order inductions to maturity of science would lead us to conclude that our science is finished, many would argue, I believe, that such inductions do justify us in considering our present beliefs well along the road to final maturity.

On what basis could such an inductive assurance of the unimportance of the inconceived be founded? Well, just as skepticism, real skepticism, about induction which refers to the vast array of evidential facts not yet considered could be at least partially muted by a consideration of the vast range and diversity of facts taken into account by our present theory, and the paucity of known phenomena as yet unexplained, skepticism of the kind we have been considering might be met by reference to the enormous richness and diversity of hypothesis types that have been considered, the extent and intensity of imagination employed in thinking up all imaginable relevant alternatives, etc. Surely there is something in the reply of the scientist to the philosopher which is often heard: "You try to induce skepticism in me by referring to the vast array of hypotheses, concordant with the data, which we haven't yet considered. But we have thought long and hard looking for such novel alternatives and have come up with none — other than the silly philosophical variants of current theory which, in the present context, just don't count."

Of course the skeptic has the obvious reply. One's certainty of the maturity of present-day science, in the sense we have been discussing, is really no more than one's certainty in the correctness (or approximate correctness) of current science. But this certainty of the irrelevance of unimagined alternatives is itself founded upon an ignorance of just what those alternatives might be. Just as we can induce skepticism with regard to current, inductively grounded, belief in science by referring to hypotheses not yet imagined, so such reference should be equally persuasive in inducing skepticism in the inductively grounded belief that, as a matter of fact, no such unimagined hypothesis would reduce confidence in our current beliefs were they to come to light.

VI

In the light of the considerations above, it is worthwhile asking what our epistemic attitude ought to be, at the present time, to our best available "cosmological" theory. Can we be assured that it is not the case that our paucity of imagination has caused us to overlook alternatives to our present fundamental physical theory which, were we to be aware of them, we would think far more plausible candidates as correct explainers of present data?

It certainly is the case that confidence in the immutability of our present theory, at least as a good conceptual approximation to the ultimate accepted hypothesis, is not universal. While many of the conceptual difficulties of relativistic quantum field theory, the closest thing we have to an overall fundamental theory at the present time, appear to be the kind of difficulties which further refinement will eliminate (I refer here to problems of divergence, renormalization, etc.), the conceptual peculiarities of the underlying general quantum formalism leave many very skeptical indeed that it, or anything quite like it, will be the ultimate way in which we will view the world. The difficulties involved in characterizing a measurement, in understanding the curious noncausal nonlocality of correlations, etc., are well known.

But at least three distinct scenarios can be offered by someone who predicts the ultimate demise of our present conceptual scheme in physics:

(1) Someone sufficiently clever will eventually see how to construct an alternative scheme which, having exactly the same experimental consequences as present-day quantum theory, lacks the conceptual elements in it which lead to difficulties in understanding. This would parallel the discovery of neo-Newtonian spacetimes as conceptually superior reconstructions of the original Newtonian theories of dynamics and gravitation.

(2) Someone sufficiently clever will eventually discover an alternative to quantum mechanics. This new theory will differ from quantum mechanics in some of its observational predictions. But the data which presently support quantum mechanics will support this theory equally well. On the other hand, since the novel alternative is so much conceptually superior to quantum mechanics, even prior to a test of the observational areas where the new theory differs from the present one, many will affirm that the new theory is a more plausible account of the old data than is quantum mechanics, and that had we seen clever enough to think it up in the first place we would have seen from the very beginning how implausible quantum mechanics was relative to this, as it happened, not then thoughtup competitor.

(3) Prior to the discovery of new data not compatible with quantum mechanics, no one will think up anything better. New observational results will, however, eventually lead to a new fundamental theory. Of course this new theory, incompatible with quantum mechanics as it may be, will account for all the observational results which made quantum theory seem so plausible before the new observational results were in. Even accepting this new theory, though, we will say that those who accepted quantum mechanics on the basis of the old data would have been justified in doing so even if this new theory had occurred to them. Just as Newton would have had a right to reject general relativity, on the basis of the claim that relative to the data then available, Newtonian theory was the simpler, and hence more plausible, explanatory account; in the light of this new theory we will still feel that, relative to the old data, quantum theory was indeed still the most plausible explanation.

Once again it is possibility (2) which interests us most here. Someone skeptical of the current theory on the basis of possibility (1) can be countered as usual with the positivist claim of equivalence of the alternatives. Someone skeptical on the basis of possibility (3) will be told that, of course, new data can always make us reject present theory in terms of a novel alternative, but that is no reason to be skeptical of the support present data gives present theory. Possibility (2) is the one which simultaneously tells us to be skeptical of even the present support quantum mechanics receives from present data, but, at the same time, gives us hope that even relative to present experimental results a diligent pursuit of imagined alternatives may very well lead us to an account of present experimental facts better than the one we presently, for want of a better alternative, accept.

Given the conceptual "nastiness" of quantum theory, one might well wish that the hopeful skeptic is right.

5. Modestly Radical Empiricism

It has sometimes been claimed that the alleged puzzles in the epistemology of spacetime and other theories are the result of a fundamentally empiricist stance toward the theory of knowledge, a stance fraught with fundamentally wrong assumptions which force upon us the apparently insoluble difficulties we face.

Two alternative fundamental epistemological approaches, pragmatism and naturalism, have been proposed, alternatives which, it is claimed by their proponents, avoid the traps empiricism gets us into. This piece examines, rather sketchily and roughly, some ways in which naturalism and pragmatism differ from empiricism with regard to three fundamental epistemic issues: the importance of the distinction between observable features of the world and those only theoretically inferred, the role of observations in providing a "foundation" to our knowledge of the world, and the role of observations in the process which allows meaning to accrue to our scientific vocabulary.

This essay takes as test cases for examining the plausibility of the empiricist, naturalist, and pragmatist models as capable of rationally reconstructing the epistemology and semantics of scientific practice two "purifications" of theory by Einstein, the elimination of the aether frame accomplished by special relativity and the elimination of global inertial frames accomplished by general relativity.

It is argued that both the naturalist and the pragmatist, in order to correctly understand these cases, will need to presume a great deal more of the empiricist apparatus of observational/nonobservational distinction than they might wish. Further, it is claimed that the naturalist or pragmatist who wishes to offer an intelligible model of theory choice in these cases will, again, be forced to come closer to the empiricist perspective of "foundational" observations than he might wish.

Finally, relying on arguments which arise when one seeks to construct an intuitively acceptable notion of theoretical equivalence, I argue that it will not be easy for the naturalist or the pragmatist to avoid components in their theory of the accrual of meaning for scientific terms which bring their theories, again, much closer to the empiricist account of meaning accrual than they might like.

I

THE observational/nonobservational distinction bears a heavy burden in empiricist approaches to theories. That such a distinction exists and that its existence is not overly context-relative seem essential if observationality is to play as distinguished a role in our account of the sources of knowledge and of meaning as the empiricist demands. Beyond this, the distinction must be "hard" enough to persuade us that the differential attitude toward meaning, epistemic warrant and (for some empiricist theories of theories) ontological commitment, which differentiates the observable from the nonobservable in empiricist accounts, can be rationalized relative to the distinction's nature. Too relative a distinction, or one that is too "soft" in other ways, will vitiate, with its modest nature, any plausibility attributable to the empiricist account.

In the realm of epistemic warrant, observations are, for the empiricist, foundational. All knowledge or reasonable belief (mathematics, logic, and the like to the side) begins with our knowledge of the contents of observation. For some empiricists observational facts can be known with certainty. For the more modest they are the only facts of the world that bear intrinsic, noninferential warrant for belief. Perhaps there are more modest empiricisms still; however, for any doctrine to be called empiricist at all there must at least be a hierarchial structure to our knowledge, with observational knowledge "closer to the base" of the pyramid of knowledge than the theoretical reasonable belief founded upon the observational basis.

Perhaps even more profoundly, observationality is, for the empiricist, the source of meaning. However meaning accrues to the theoretical vocabulary characterizing nonobservable entities and properties, it must first accrue to language describing the observational features of the world; only then can it percolate upward to the remaining portions of discourse. However modestly and delicately construed, some form of intending to use a word, accomplished by means of ostensive definition, must be at the root of our possibility of meaningful discourse. Whatever empiricism is, it entails the view that there is a world external to and independent of the language we use to describe it and that the junction point between language and world is located in the presence of the world to the language user in observationality.

But, of course, all this has come under merciless assault in recent years, having been subjected to arguments (some new and some revisions and modernizations of the old) intended to show that the whole empiricist program rests on a collection of mistakes, howlers, and nonsequiturs. We are told that any possible distinction between observable and nonobservable is context-relative, merely pragmatic, soft, variable, and in any case nothing like the kind of distinction needed to support the empiricists' radically different construals of claims about the observable and the unobservable.

Further, we are informed, even if such a significant distinction between the observable and the nonobservable were possible, the observable could in no way play the significant roles reserved for it by the empiricists. Epistemically, foundationalism as a whole is simply wrongheaded. Knowledge is a web, a network of belief. There is no bottom, no top, and indeed no hierarchical direction at all. So there is no need for observational beliefs to provide a foundation on which the whole structure of knowledge rests — which is a good thing, as they could not possibly play such a role were a candidate for the job needed.

Semantically, too, we are told, the empiricists' claim of a special role for observations in the grounding of meaning is a myth. Language is public, and meaning is use. If there is anything left for meaning to be at all, it is something to be approached from the viewpoint of a holistic, functionalistic account of communication. There is no place left in any serious meaning theory for a special role of observations, functioning through ostensive definitions, to provide the basis of meaning accrual — again, a good thing, since Wittgenstein has shown us that observations in the empiricist sense, even if they existed, simply could not play the role demanded of them by the empiricist account of meaning.

At least two related but quite distinct accounts of knowledge and meaning are offered to take the place of the discredited empiricist account: naturalism and pragmatism. I intend to sketch, in the broadest and crudest possible strokes, some of the aspects of these two alternative accounts of knowledge and meaning. Then I plan to look at the epistemic and semantic role that observationality seems to play in a number of apparently special theoretical contexts. I will argue that neither the naturalistic nor the pragmatic account of the role of observationality does justice to the special role which the observational/nonobservational distinction plays in these special theoretical contexts, and that these two alternative accounts also fail to do justice to the apparently very special place of the observational in epistemic and semantic critique of theories (again in these special contexts). I will argue that the empiricist account, whatever its ultimate faults, seems to do a better job of reconstructing and rationalizing our actual scientific practice in these special cases. Then I will try to explore a little the question of just what makes these cases special and therefore supportive of empiricism against its rivals. In particular I will ask whether the support given to the empiricist account in these cases can be generalized to argue that, overall, empiricism just seems to do a better job than its rivals of accounting for facts and intuitions.

II

The naturalist tells us to abandon "armchair apriorism" in philosophy. If there is anything we can reasonably say about the world and our place in it, it is only that which is a consequence of our best currently available theory of the world. This theory, which summarizes our best established scientific knowledge of the world, is the repository of our knowledge of the world and tells us all there is to know. Nothing comes before the theory and nothing stands outside it. If epistemic critique and semantic analysis are to be done at all, they are to be done as merely specialized components of the general scientific attempt to grasp the nature of the world.

The theory, according to the naturalist, determines what is observable and what is not. Observers and their observations are as much in the world as tables, clouds, and quarks, and if there is anything to say about them our theory of the world says it. This, of course, could be compatible with many positions on the observable/nonobservable distinction. But, in fact, the naturalist camp has a fairly unified view. The theory of the world is physicalist. Observers are more or less reliable measuring instruments, responding to stimuli impinging on them from the remainder of the world in causally correlated ways, allowing us to infer, more or less reliably, from their functionally characterized states to states of impinging environment. Of course, from this point of view the observational/nonobservational distinction is highly context-dependent and relative. Functionally characterized brain states are "observations" relative to the psychologist's interest in experiments on perception; however, relative to the astronomer's interest in stellar constitution, spots on photographic plates or the output of image-amplifier tubes function just as well as the "observations" correlative to the particular phenomena being observed.

Knowledge, from this naturalist perspective, is just true belief acquired by a reasonably reliable process. The only grain of truth remaining of the empiricist view that observation is at the ground or root of all knowledge is the empirically established truth that, as a reliable indicator of what the world is like, what we take to be observations (in the naturalistic sense of how observations actually function in science) are a more reliable indicator of the true state of the world than are, for example, wild guesses or the intuitive states of the seer. This fact of reliability can, of course, be explained by the causal origin of observations correlating them to the states of the impinging world. Epistemic critique, from this point of view, can only amount to demanding of a theory of the world internal self-consistency. Something is wrong if the theory tells one, in its own terms, that one's grounds for believing it are unreliable. Beyond this one cannot go.

About meaning we have much less in the way of firm agreement from the naturalists, but all naturalists agree that an account of meaning is to be in terms of some sort of functionalist account superimposed on an underlying physicalist world view. The role of language in public communication, and meaning as some sort of "coarse grid over use," are common elements of the sketchy naturalistic semantics we have been offered. Although constraints of rationality may leave a semantic account of meaning distinct in kind from the physicalist lawlike basis that describes the world of things and events from a naturalistic point of view, one thing is clear: There is no place for some sort of private first-person intending of words to things by the mediation of ostensive association of words with observational content in any kind of respectable, naturalistic meaning theory.

It is harder to pin down the pragmatic approach than the naturalist, even in the crude way we must be satisfied with here. Essential, surely, is the denial of any kind of viewpoint external to our immersion in our present schema of belief from which the schema as a whole could be epistemically or semantically examined and criticized. There is only the ongoing process of piecemeal revision, guided by precepts of rejection and innovation, themselves the product rather than the presupposition of our accepted ways of acting in the world, where action includes as but one component the activities of speaking and describing.

Whatever the distinction between observable and nonobservable may be for the pragmatist, it is most assuredly context-dependent and relative. We speak correctly of persons observing the red patch before them (or even appearing to them in hallucinatory experience), but equally correctly of the physicist observing the passage of the K meson in the Wilson cloud chamber. Of course, our distinction between what we observe and what we infer may sometimes have a point, say in offering an appropriate in-context critique of someone's illegitimate assurance. (A defense attorney asks a prosecution witness: "But you didn't actually observe the murder, did you? You were on the other side of the door." However, hearing the screams of the victim from the other side of the door certainly is "observing the murder" when contrasted with the experience of the murder being gained by the jurors in the courtroom.)

What is the epistemic special place of observation for the pragmatist? For the pragmatist, what is desirable is parasitic on what is desired. Similarly, what is believable is parasitic on what is believed. Our rules for rational believability are simply the reflective synopses of the principles by which we do in fact accept beliefs as warranted. From this point of view observations do have a kind of epistemic priority, but it is nothing like that given to them by the empiricist. Since observations do, in general, constitute the locus of firm, shared agreement, and since agreed belief is the only ground on which agreed rational believability can be based, the fact that beliefs accrued by observation are usually beliefs shared with some degree of certainty places them in a more or less central point in our structure of accepted truth. But that is all there is to their centrality.

Semantically, with the pragmatist as with the naturalist, observations are reduced to a much lower role than that granted them by the empiricist. Meaning, if it is anything at all, is use. A notion of meaning may play a role in some discourse used to talk (more or less theoretically) about discourse, but no special place will be played in that account by any mythical act of intending a word to designate an item of "pure experience." A *reductio ad absurdum* of the empiricist view is provided, from this standpoint, by mere reflection upon the fact that before experience can enter into our cognitive structure of belief it must be experience "always already" conceptualized and linguified. All terms function according to their total role in our language scheme for finding our way about in the world; all discourse is "always already" theoretical discourse, and all experience is "always already" framed in terms of our conceptual apparatus with its embodied theoretical presuppositions. Of course, the answer to the empiricist's question as to how then language connects up with a preconceptualized world is that the very idea of such a world is a well-lost myth, part of the naively dualist scheme of world and representing perceiver.

III

As I noted earlier, my aim is not to pursue in extensive and careful detail just what the empiricist, naturalist, and pragmatist options are, but rather to look at their suitability, crudely characterized as they may be, to give a rational reconstruction of our actual scientific practice in a couple of apparently very special cases of scientific theorizing. Later we shall consider whether these cases are really as special as they first appear to be, or whether instead they ought to be taken as paradigmatic of the case of rational scientific decision-making, differing from the more usual cases only in having presented in the special cases in a bold and surface way certain features that are only more subtly discernible in the more standard cases.

The special cases I have in mind are Einstein's two "purifications" of physics: the purification by elimination of the aether frame, which is the fundamental accomplishment of special relativity, and the purification by the elimination of global inertial reference frames, which is the fundamental accomplishment of general relativity.

In the first case there was an antecedent theory, Maxwellian electromagnetism, which as Einstein clearly saw, possessed an element usually otiose for the prediction of observable phenomena: the absolute velocity of the system relative to the aether frame of electromagnetism. The null results of the famous round-trip experiments on light eliminated what could plausibly be construed as the last possibility for finding some observational consequence of uniform motion relative to the aether frame. Einstein's genius allowed us to see that, with a sufficiently radical revision of the very most fundamental ideas of space and time, a new electromagnetic theory could be constructed that would give the usual observational results in the standard experimental cases and the null results in the case of the new round-trip experiments, without the peculiar "putting absolute velocity in only to take it out again" that infected the earlier compensatory theories. Of course he then went on to invent a novel mechanics properly invariant relative to the new spacetime picture, but this, important as it is to physics, is for us only a derivative consequence of the fundamental move.

The origin of general relativity can be viewed from a similar perspective. The principle of equivalence shows us that the traditional joint theory of mechanics plus gravitation posits a distinction without an observational difference. A world with a uniform gravitational field everywhere is indistinguishable by any observational means from one without that field. The theoretical otioseness can be cashed out by moving from a theory that treats the global inertial frames of spacetime and the gravitational field separately to a new pseudo-Riemannian spacetime in which nothing but the geodesic structure of the spacetime exists, determining the local inertial reference frames and simultaneously fixing the inertial and gravitational properties of the world everywhere. Once again there is much left out in this picture (such as the fact that it is now relativistic gravitation that is assimilated to spacetime structure and not Newtonian gravitation, and the importance of the generation of the new field equations connecting geodesic structure to mass-energy source) but for our purposes these are again not nearly so interesting as the fundamental move of purifying an antecedent theory of elements that are totally otiose from an observational point of view.

What is special about these cases of the construction and the acceptance of scientific theories? For one thing, the distinction between the observable and the nonobservable is crucial to the whole scientific enterprise in these cases. Unless we are assured that in our notion of the observable consequences of a theory we have genuinely captured all possible observables, the reasoning just does not go through. In both special relativity and general relativity we are told to delimit ourselves to coincidences and continuity along paths traversible by causal signals in trying to map out the space-time structure of the world. If we allow ourselves to go beyond this (countenancing, say, simultaneity for spatially separated events, or the absolute magnitude of the gravitational force field, or the geodesic structure of spacetime itself among the class of things that could, in principle, be observable), then the whole Einsteinian program of demonstrating to us the theoretical otioseness of aspects of established theory, and then revising theory in such a way as to eliminate from it those elements totally irrelevant to observational prediction, becomes an impossibility. If we could ever, in principle, determine simultaneity for distant events in a noninferential, observational way, then all of the arguments designed to show us that the aether frame is an in-principle otiose element of physics fail. And the argument goes through in the same way regarding the allegedly otiose notion of a global inertial frame dispensed with in general relativity.

This is not to say that the features of the world called "observable" in presentations of relativity are what the radical empiricist philosopher would count as really, truly, genuinely observable; only that what is counted in these theories as forever unobservable must be, in principle, in that class. For Einstein's arguments to be the least bit persuasive we must be assured, in a manner independent of the theories proposed and ultimately accepted, that there are nonrelative, noncontextual limits on the domain of what is to be counted as observable. This holds true independent of one's views of the status of the theories one ultimately accepts in an Einsteinian way. Whether one believes that the theories are to be interpreted realistically or instrumentalistically, and whether one believes that the choice of alternative theory is a matter of convention, of a priori plausiblity, or of projection from past accepted theory, the very structuring of the scientific decision process requires that one not be prepared to challenge Einstein's hard, context-independent, and irrevocable limitation on the domain of the observable.

Other aspects of this problem of theory choice have a degree of "specialness" as well. The theories presented (Minkowski spacetime, curved spacetime) are novel and radical. Whereas the older theories they are to replace have a certain familiarity and built-inness as components of our ordinary way of looking at the world, the new alternatives—the ones any rational person would accept according to most of us with intuitions about this—present us with a view of the world sufficiently startling that it takes us a lot of effort and use of analogy to feel that we really even understand exactly what the picture of the world presented to us by the new theory amounts to.

Furthermore, we have in these cases a presentation in a fully worked out form of something philosophers frequently allege to exist in other cases but have a hard time demonstrating. We have alternative, fully characterized and developed, theories of the world, all of which agree with each other on what has been presupposed in the context to exhaust all possible observational data. It is one thing to be told that life might be a dream or that one might be a brain in a vat and yet all one's experiences might be the same; it is another thing again to have two fully developed theories before one, and a fully characterized notion of what constitutes all of their consequences plausibly characterized as observational, along with a demonstration in the formally fullest manner that, as incompatible on the surface as the theories might be at the nonobservational level, their observational consequences are provably the same.

Nor, in this special case, can any of the observationally equivalent alternatives be glibly dismissed as so intrinsically implausible as to be unworthy of serious scientific consideration. Perhaps we can dismiss dream-life or brain-in-a-vat hypotheses as beyond plausibility—mere philosopher's alternatives that no rational man in the course of framing a theory about the world would ever seriously consider. But in the special cases we have been considering such instant dismissal of alternatives would plainly be out of order. The alternative accounts to those we do accept (special and general relativity) not only are viable scientific hypotheses, they are the hypotheses all reasonable people in the scientific community did hold to (sometimes only implicitly and without even noticing that they were hypotheses) before Einstein pointed out to them that new and better alternative accounts of the world were available.

IV

How do the claims of the naturalist fare in an attempt to understand what is going on in the special scientific situations on which we have focused our attention?

First there is the naturalist's claim that any relevant observational/nonobservational distinction must be merely part of an established naturalistic picture of the world, and that, seen as such, it can be clearly seen to be at best context-relative. The theory tell us what is observable. But does it in these cases? I doubt it. Notice the independence of the assumption that the observational facts grounding spacetime theories must needs be local facts and facts about material systems. Coincidence of events and continuity of the path of a particle are legitimate components of observationality. Global facts such as simultaneity for distant events or facts about the structure of the spacetime itself (whether its geodesics are straight lines) are not even considered as possible components of the observable world. Where do these assumptions that must be made for the whole critique to get under way come from? From a naturalistic investigation of the actual constitution of observers as components of the naturalistic world? From a physicalist examination of their structure as more or less reliable recording instruments responding causally to the data? I doubt it. There is built into this scientific practice an aprioricity of what is observable that, whatever its origin, just does not seem to fit the naturalist's model.

This can be seen clearly when we reflect on the fact that observationality as it is meant in these critiques just cannot be taken to be the context-relative thing the naturalist demands it to be. To be sure, once we have adopted general relativity we can speak of the astronomer "observing the null geodesic structure of the world" by watching the course of unimpeded light rays in outer space. That notion of observing is plainly not the same one which we needed to get general relativity in the first place, though, for if we could have, all along, observed the geodesic structure of spacetime, we could have told from the beginning whether we were inertial or, instead, noninertial but falling along with everything else in a uniform gravitational field, and we would have never dropped the global inertial frames and gravity in favor of local inertiality (that is, in favor of general relativity) in the first place. The critiques Einstein gave us, however they are to be read, seem to require an absolute notion of at least nonobservability to get them under way, a distinction between the really nonobservable and the observable quite unlike anything compassable within the purely naturalistic scheme.

Epistemically, the naturalist tells us that all we can hope for is an indication that a method of inference is, according to the theory of the world that we do accept, a reliable method for coming to the truth. We know, a priori, that such a notion of evaluating a method will be of no help to us when we are puzzled about just which of a number of seemingly equally good theories of the world we ought to believe, where the theories are not minor hypotheses about things whose general nature we already understand but radically distinguished general accounts of the structure of the world.

Given the data behind special relativity, ought I to believe that theory or perhaps one of the aether-theory alternatives? The naturalist says, "Believe the theory that is determined to be believable by the methodology that will, according to your best theory of the world, most often lead to the truth." But how is this any help at all? If the world really is as Einstein tells us, then the usual procedure that leads us to special relativity is the one that is reliable. If the world is, rather, as the aether theory tells us, then the usual method is not reliable, for it misled us in one of the fundamental cases in which it was applied. But, of course, what we want to know is which alternative we ought to believe, sitting here with the data and with general inductive good practice and dubious about the fundamental facts of spacetime.

Lots of options do occur to us that have the virtue of being relevant to this situation. We can opt for skepticism and withhold judgment. We can opt for simplicity or conservatism or a priori plausibility to resolve the dilemma. We can develop a confirmation theory that gives the alternatives differing degrees of credence, even relative to exactly the same observational confirmations. Or we can try to undercut the skeptical problem in the traditional positivistic manner of arguing that there is no theory choice to be made at all, since observationally equivalent theories are fully the same theory differently expressed.

None of these may be satisfying solutions, but they are at least relevant options to be proposed here. That we, given our theory of the world, externally evaluate others as more or less reliable in their means and methods for discerning what is or is not the case is true enough, but it is totally irrelevant to the internalist problem of trying to extrapolate from what we can observationally discern to what the world is like. All of this is clear enough in an appraisal of reliabilism from a general perspective. However, faced with the problem of real, fully formulated alternatives, all of which coincide on the presupposed and generally unquestioned basis of what is taken in the scientific context to exhaust the observable, the criticism of reliabilism becomes at least more persuasive and more dramatic.

The naturalist does, of course, have a reply to this empiricist critique of his position as incapable of rationalizing the inference from observable data to inferred theory. The line we would expect would go something like this: If we can establish, on the basis of our best accepted theory to date, the reliability of some principle allowing us to infer from data to generalization, then we are justified in using that rule to project from new data to new generalizations. For example, relying on our accepted paleontological-geological theory of the world, we are justified in inferring from a newly discovered fossil type to the existence of a previously unsuspected extinct species rather than to some curious creationist alternative. But how would such a reliabilist rationalization of projective inference work in the cases in question? Only, I think, at a rather high-powered metalevel. If our best accepted theory to date could tell us, for example, that simple theories of the world were overwhelmingly more often true than their more complex alternatives, then we could, on this naturalistic ground, rationalize the choice, say, of special relativity against the aether theories. I am extremely dubious, however, that any kind of reliabilist account of grounding the usual principles that allow us in the broadest cases to make our leap beyond data to theory can be made plausible. Where we are making limited inferences from only somewhat novel data to only somewhat novel new theories that are small components of our overall world view (as in the paleontological case noted above), we can have some hope that a broader component part of our accepted theory of the world can serve to rationalize our theory choice. Where, however, we are making global inferences from broad reaches of fundamental data to theories that themselves define our overall world picture, and where we have little to go on other than the traditional notions of simplicity of a priori plausibility in making our choice, I doubt that any attempt to characterize the rationalization of this kind of theory choice as being of the reliabilist type can succeed.

What about the naturalistic approach to semantics? How well does it fit in with what we seem to be impelled to say about meaning when dealing with the cases to which our attention has turned?

It is harder to say anything definite here than it was in the case of naturalistic epistemics, but I think that the general naturalistic approach tends to leave out something of crucial importance when it deals with cases of this special kind.

To say that meaning is derivative from use is vague and harmless enough. Where I doubt that the usual naturalistic approach to semantics functions adequately in our problem cases is where it disavows any special role in a meaning theory for ostensive association of words with directly presented situations of the world.

We need meaning theory in the problem cases primarily to resolve the question of the equivalence or the nonequivalence of observationally equivalent theories. I need not go into detail concerning all the options available here (Ramseyite approaches to theoretical terms, semantic analogy approaches, Craigian equivalence doctrines about theories, and so on) to say at least this: Any resolution of the problem of the standards for equivalence of theories seems, at this point, to require a radically asymmetric treatment of terms that are taken as referring to nonobservable entities and properties and terms that function by referring to the observable.

Not only the positivist, but even the realist, who wishes to have some standard for determining the equivalence or the nonequivalence of theories, seems to require that a certain part of the vocabulary of science be distinguished from the remainder in an adequate treatment of the accrual of meaning. Whether one believes the aether theories and special relativity to be equivalent (because of their observational equivalence) or to be nonequivalent (because of their nonisomorphism on the theoretical level) despite their observational equivalence, one will find oneself, when one takes the problem of equivalence seriously, treating the terms of the presupposed observational language radically differently from those of the nonobservational discourse. However it is done, sooner or later one finds oneself treating the theoretical discourse as parasitic on the observational. For some radical positivists the former is simply termwise definable by the latter. For some realists the theoretical terms are holistically fixed by their Ramsey sentence role in the total theory. Both approaches, however, require that the observational discourse come fully equipped in a meaningful way prior to its functioning in the theories in question.

Perhaps a relativity is implicit here. Vocabulary that is observational relative to one problematic will function as theoretical in another context. Perhaps there is no ground level of irreducibly observational discourse. Nonetheless, I think that ultimately one will have to allow for some discourse a distinguished role, in that its meaning is fixed not by role-in-theory but by some version of the ostensive intending of word to world mediated by presentation of the world in awareness so familiar from the empiricist tradition and so fervently eschewed by most naturalists.

V

How well does the pragmatist approach to the notion of observationality and to the epistemic and semantic role of observations fare when brought to bear on the special cases to which we are attending? Once again, I think that careful examination will show that empiricism, whatever its faults, gets us closer to what is going on than does pragmatism.

The pragmatist, perhaps even more fervently than the naturalist, wishes to deny the existence of any firm, context-independent distinction between the observable and the nonobservable. As before, I do not think this does justice to what goes on in the Einsteinian critiques. Once again, the claim is not that what is taken as observable in these critiques constitutes the ultimate level of pure, noninferential observationality, or even that such a level exists. Rather, the claim is that what is counted as nonobservational in the Einsteinian critiques must, for these critiques to be even remotely plausible, be nonobservational in some strong, in principle, nonrelativized way. Were it even conceivable to us that we could, in the real sense of observability, observe whether distant events were simultaneous or observe the geodesic structure of spacetime itself, then the demonstration on the in-principle otioseness of the earlier theories would not go through. I do not believe that one does full justice to Einstein by viewing his alternatives, special and general relativity, as simply more plausible alternatives than the aether and spacetime-plus-gravitation theories they replace. What is crucial to Einstein's accomplishment is his demonstration of the methodological deficiency of the earlier theories. They simply are not the right kind of doctrines to be even considered legitimate candidates for the correct explanation of the world. Einstein's argument requires a conviction that some aspects of the world in these earlier theories' ontologies are unobservable in a rigorous, non-context-dependent way. This is not to deny, for example, that we are then permitted to talk, in the loose sense, of astronomers observing the curvature of spacetime.

Consider, in addition, the pragmatist claim that all observations are theory-laden. Perhaps they are, but the whole point of the Einsteinian critique, which emphasizes the restrictive observational basis we can truly count on in constructing spacetime theories, is to show us that with careful reflection we can, in a specific context, purge our characterization of the observational data of those theoretical presuppositions that stand in the way of our discerning what is truly empirically relevant in our theories and what is conceptually otiose. The whole point of the critiques, with their emphasis on coincidence and continuity along traversable spacetime paths as the limited body of truly observational data available to us, is to get us to see that, once purged of theoretical preconceptions, our data can be captured by a theoretical apparatus that is far more parsimonious in its ontology than were the earlier theories and is superior to them in that what has been discarded ought never to have been tolerated by a respectable methodology anyway.

Does this mean that we can eventually discover a basis of observationality purged of all theoretical preconceptions? Perhaps not. It is not so radical a radical empiricism that I am espousing. The important thing about the empiricist claim, which the pragmatist denies and which must be invoked to make sense of the Einsteinian critiques, is the hierarchical nature of our science, in which, lower in the hierarchy, are observations that in a given theoretical decision-making context have been detheorized in such a way as to purge them, as observations, of theoretical presuppositionality that would get in the way of a fair epistemic evaluation of the contending theories. Once again, once we have opted for one theoretical approach, our language of observationality may become reinfected with theoretical presupposition. Even our immediate data of private awareness may become structured, in the familiar Gestaltist way, with our theoretical bias. But it is the possibility of a rigorous purging of the relevant theorization of observational data that is relevant, and that is just the sort of thing Einstein shows us how to carry out in these particular cases.

Does the pragmatist do justice to the epistemic place of observations in these cases? I think not. Remember that for the pragmatist the only things that observation statements have in favor of their special epistemic role is that they are a class of usually agreed upon, uncontroversial truths. I think that misrepresents the case, however. I think it is more plausible to claim that our naive spacetime picture garnered far more instinctive agreement among the community than would the truth of the null results of the Michelson-Morley experiment. Nonetheless, the latter take dominance over the former when theoretical decisions are made. Theories, as well entrenched as we can imagine a theory to be, must make way in the face of even a few novel bits of contradictory observational data. Of course there will be cases where we are so convinced of the truth of a theory that we will not take alleged data contradicting it very seriously. The response of the physics community to Miller's later alleged positive results of round-trip experiments shows us that. Yet there is still a fundamental sense in which Popper is right that, when the issue is between what a theory predicts and what the observational results tells us is (contrary to the theory's prediction) really the case, it is observations which must take epistemic precedence if there is to be anything like a coherently rational science at all. Does this mean that there are incorrigible, indubitable, or irrefutable reports of observation, or anything like that? No. Once again the fundamental point of empiricism is not that "foundations" exist but that there is a hierarchical order to our epistemic structure in which observation is foundational relative to theory. This is what the thoroughgoing pragmatist coherentist denies.

When we look at how we get from data to theory, what these special cases tell us about pragmatism (as opposed to empiricism) must be construed more subtly.

Some basically empiricistically minded philosophers, looking at the underdetermination of theory by data, that is clearly present in the cases we are considering, opt for a theory of theory choice that invokes such notions as ontological simplicity and methodological conservatism, to allow the selection as "best" of one of the many alternatives, all of which are equally compatible with all possible observational results. Surely, if the choice is done that way we are well on the road to some kind of pragmatist epistemology. How else, other than by the reduction of truth to ultimate warranted assertability and the reduction of assertability to the systematization of what we in fact take to be warrantedly assertible, could the connection between simplicity or conservatism and truth ever be motivated?

However, deeper reflection on semantic issues will show us, I believe, that what these special cases really reveal to us is the necessary absorption into pragmatist accounts of meaning of very significant portions of the empiricist account. A standard complaint lodged against pragmatism, with its coherentist notions of warrant and truth, is that too many incompatible worlds could all be the case on the pragmatist view. Recently we have been assured, on the basis of arguments stemming from the problem of radical translation and the inevitable role in it of the principle of charity which attributes to others a general agreement with us to the facts of the world, that we need not worry about such "alternative coherent worlds" ever arising. Given a coherent account of the world, we will perforce interpret it as, in general, the same account of the world that we offer in our home world picture.

It has been claimed that this pragmatist solution to the problem of too many coherent world views is basically verificationist. Indeed, I think it is. Rather than go into that, all I want to do here is show how reflection on the cases at hand will indicate how a careful construal of pragmatist attitudes toward semantics may allow us to see that at least some pragmatists are more empiricist than they would like to think themselves.

Just as the recent pragmatists evade the issue of multiple coherent worlds, some have tried to avoid the problem of theory choice among spacetime theories with all observational consequences in common by arguing that there is no choice to be made. All the observationally equivalent alternatives are said to be mere alternative expressions of one and the same theory. I think it is clear that the only way that this line could be made to work, in general, would be on the basis of a positivist view that what a theory says is just what is said by its totality of observational consequences. Surely that requires the full empiricist panoply of hard observational/ nonobservational distinction and the very special role of observations in grounding meaning which are the hallmarks of empiricist semantics.

Actually, I think much more is true. Even someone who wishes to claim that aether theory and special relativity are not equivalent theories, and then to be a skeptic, perhaps, or else to invoke some principle above and beyond conformity with the data that rationalizes theory choice, must still offer us a coherent account of just what makes theory expressions expressions of one and the same theory (when they are) and inequivalent expressions of distinct theories (when they are not). I believe that one really cannot give such an account in any coherent way without presupposing a good deal of the empiricist's apparatus. In particular, I do not think that, without a rather rigid distinction between the observable and the non-observable (again, a distinction at least rigid enough to assure us in any context that a body of propositions contains all that could be plausibly construed as observable, assuring us that all in the re-mainder is nonobservable in a context-independent way), we can construct any plausible general principles of theoretical equiva-lence, whether these principles (as in positivism) construe aether theory and special relativity as equivalent or whether (as in some Ramsey sentence approaches), they construe them as nonequiva-lent.

Generalizing, I believe the following to be true: Any pragmatist who wishes to avoid the familiar accusation that his principles of epistemic warrant will allow as warranted (and, from a pragmatist viewpoint as "true") too many alternative internally coherent worlds will have to come up with a notion of equivalence of world pictures that parallels the positivist picture of equivalence for theories applied in the cases we have been discussing. To this ex-tent, such a pragmatist will be presupposing, in a manner very distasteful to him, the empiricist notion of genuinely nonobserv-able aspects of the world and the empiricist idea that observational facts play an important and distinguished role as the "ground" of the accrual of meaning to propositions about the world.

VI

In summary, I am arguing that neither naturalism nor pragmatism does justice to the nature or role of observations in the context of Einstein's critiques.

Both pragmatism and naturalism advocate a merely relative and contextual distinction between the observable and the nonobserv-able. However, Einstein's critical results do not fully make sense unless at least some consequences of theories can be taken by us to be, in principle, nonobservational in a nonrelative and non-con-text-dependent way. In addition, naturalism, with its doctrine that what is observable is determined by the theory, neglects a priori aspects of the observable/nonobservable distinction which allow us to make that distinction prior to accepting a theory to cover the data and independent of the theory we will ultimately adopt. Pragmatism, with its doctrine that observations are all intrinsically theory laden and that facts are "soft all the way down," fails to do justice to the way in which someone like Einstein deliberately shows us how to at least partially (and relevantly for his purpose) detheorize our observations so that they can be conceptualized in a manner independent of the theories from among which we are making our choice of which to believe.

Neither naturalism nor pragmatism does justice to the epistemic role of observations in the theoretical contexts in question. Naturalism restricts us to asking questions about the reliability of methods, evaluated externally and relative to an adopted theory of the world. What we need to attain in the scientific case in question is, however, an understanding of the internal role observations play in rationalizing for the undecided decision-maker himself, one of his possibilities as the rational theory to believe. Pragmatism, which can deal with the special epistemic role of observations only by characterizing them as the sorts of things commonly agreed upon by the members of the community without doubt or disagreement, fails to do justice to the quite asymmetric role played by observations (no matter how dubious) and theories (no matter how entrenched and agreed-upon) in the process of scientific decision-making.

Nor can naturalism or pragmatism do justice to the fundamental semantic role played by observationality in the contexts in question. Both approaches, each locked into one or another use theory of meaning, fail to provide a radically distinctive role for observables as opposed to nonobservables in an account of meaning. Resolution of the scientific problems we have been discussing requires analysis of the notion of equivalence of theories, however. Whether one adopts a positivistic account in which any observationally equivalent theories are taken to be fully equivalent or instead an account that allows for observational equivalence without full equivalence (say, by demanding structural isomorphism at the theoretical level for full equivalence), it can still be shown that observational identity is crucial for theoretical equivalence and that, as a consequence, radically distinct accounts must be given in one's theory of meaning to the accrual of meaning by observational terms and the accrual of meaning for the terms referring to the nonobservable entities and properties. Although a naturalistic-pragmatic use account of the latter, relative to an assumed meaningfulness already accruing to the former, may be plausible, such an account will not do for the observational part of the language.

VII

The fundamental claim being made here is this: The first-person, internalist perspective is an essential component of any thoroughgoing epistemic and semantic critique of theories. It is indispensable and cannot be replaced without loss by either a physicalistic-naturalistic epistemology and semantics that has been naturalized or a pragmatist "always-already-ism" or internal coherentism with respect to epistemic grounding or semantic comprehensibility.

The perspective I have been emphasizing is a hierarchical one. With regard to epistemic assurance and to semantic comprehension, some things are more foundational than others. The existence of some ultimate foundation, some bottom to the hierarchy, is not of the most fundamental importance. What is essential is that there is a "closer-to" and a "further-away from" the foundation, and that this directionality is provided by a measure of the degree to which we have moved away from the observational basis of theory. What is important is not whether we can ever find a kind of assertion that, being purely a report of the immediately observable, is immune to epistemic doubt and has its meaning given by some infallible ostensive process. What is crucial is that we can develop an epistemic and semantic critique of theory that proceeds by moving to a relatively purer epistemic and semantic level. In addition, it is crucial that this procedure is the process of extracting from the theoretical context that which is to count as more given and less theory-laden —i.e., that which is to count, relative to the context, as the observational basis of theoretical decision-making as opposed to theoretical presumption.

Now, it might be thought that such a line of reasoning appears plausible only because of the very special nature of the theorizing context from which our examples of scientific practice have been chosen. These contexts are indeed special in some ways. The theory choices (aether vs. special relativity, gravity vs. spacetime curvature) are totally formulable, unlike the vague and unformalizable options we talk about in the more philosophical context (such as material world vs. brain in a vat). We are in a situation where a genuine theory choice must be made. We cannot rely on everyone simply agreeing a priori that all but one of the alternative theories before us may simply be dismissed as philosophers' nightmares, that are not worthy of serious consideration in practice. The epistemic-semantic critique of the theories is, in the context in which we are working, an essential part of the very formulation of the theories in question. In the development of special and general relativity, the investigation into what is and what is not observable and the role of that distinction in characterizing the epistemic and semantic bases of the theories were essential to getting Einstein's practicing science under way. Here the epistemic-semantic critique is integral to the ongoing scientific program, not a philosophical afterthought. In the cases in question, the distinction between what is to count as observational (coincidence, continuity along timelike paths) and what is nonobservational (nonlocal spatiotemporal relations, the structure of the spacetime itself) is clear and uniformly presupposed as a given by the scientific community.

Granted, all of this is rare in the scientific situation. Usually science goes on without such an epistemic-semantic critique, leaving that for philosophical "moppers up." But I think the ways in which the particular problem situation focused on here are special make it clear that, in principle, all our commonsense and scientific beliefs ought to be susceptible to the same sort of epistemic-semantic critique. That this is so, that in any such critique reliance on a first-person, internalist perspective is essential, and that from that perspective both epistemological grounding and semantic comprehensibility require a special privileged role for the observable are I think, at the core of what empiricism is all about.

6. Inertia, Gravitation, and Metaphysics

The focus of this chapter is on the relationship between the scientific contents of a theory and the "metaphysical" implications we may draw from it. The scientific theories taken as illustrative are those which replace flat spacetime structures equipped with gravitational fields by curved spacetimes in which gravitation has been integrated into the spacetime geometry. The metaphysical issue is that of realism vs. instrumentalism in the interpretation of theories.

Beginning with puzzles encountered in Newtonian cosmology, I outline the way in which one might progress from a flat spacetime plus gravitation to a curved spacetime theory, along with the similarities and differences in the Newtonian (prerelativistic) and the relativistic cases.

Next I examine the epistemic motivations and metaphysical presuppositions which might lead one to suggest that the curved spacetime theory is a better alternative than the empirically equivalent flat spacetime alternatives. I then try to show that a reasoning not altogether unlike that which leads us from flat to curved spacetime views of the world might lead us further to an antirealist attitude toward spacetime as an entity altogether.

It is argued, not that either realism or antirealism is the correct attitude to adopt toward spacetime theories, but only that the proponent of a realist attitude toward curved spacetime must face up to the obligation of explaining to us why the methodological principles which led him to curved spacetime in the first place do not lead further to a relationist denial of spacetime as an entity altogether.

I

ATTEMPTS at deducing empirical science from metaphysical postulates alone are numerous and well known. So are their failures. But one can work the other way and try to derive one's

I am grateful to the anonymous referee for *Philosophy of Science* for helpful comments on the original version of this paper.

metaphysics from one's "best available scientific theory." Given the difficulty (impossibility?) of ever distinguishing "metaphysical viewpoints" from "physical hypotheses," this program is, in one way, unassailable. Simply argue that a given metaphysical view is the "best supported explanatory hypothesis" in some area of science, and then deduce the metaphysical view from itself.

But there is a more honest approach, and one which would be infinitely more fruitful were it to succeed. Pick some intuitively, patently metaphysical position and some intuitively, unquestionably scientific hypothesis. Try then to show that the latter at least "supports" the former, even if the former cannot be "deduced" from the latter in any straightforward sense.

Newton tried this approach, seeking to establish the metaphysical doctrine of the substantiviality of space on the basis of simple, mechanical experiments and the theory necessary to account for them scientifically. Latter-day "Newtonians" try the same approach. Only now it is *spacetime,* indeed curved spacetime, which is the "entity" whose existence is to be demonstrated, and it is general relativity, not Newtonian mechanics, which is to serve as the scientific support of the metaphysical position.

I will argue here that the inference from science to metaphysics of these modern "Newtonians" relies upon inexplicit, presupposed assumptions which are, intuitively, just as "metaphysical" as the doctrines they are used to support. So if the game is played honestly what the "Newtonians" show is not that metaphysics can be inferred from science, but that from science and presupposed metaphysics one can come to metaphysical conclusions. Or, alternatively, metaphysics will follow from science if you build enough of the former into the latter to begin with. This, I submit, is a far more modest achievement than that which some "Newtonians" would like us to believe they have accomplished.

In the course of pursuing this major theme I will also have occasion to touch, once again, upon some aspects of the perennial problem of the alleged "conventionality" of theories about the structure of space and time.

II

Let us begin with a straightforwardly scientific problem: the construction of a Newtonian cosmology. It was already realized in the nineteenth century, by Neumann and Seeliger, that the construction of a model for the universe which was infinite in extent, and which had a nonvanishing mass density everywhere, involved great difficulties—especially if one had in mind a static model.[1]

Following the development of the first general relativistic cosmologies the Newtonian problem was approached again. Milne proposed a Newtonian model of a universe containing an isotropic, expanding matter of uniform density which had close analogies with the Einstein–de Sitter model of general relativity.[2] This was generalized by Milne and McCrea to provide Newtonian models analogous to the general class of Friedmann models studied in general relativity, including models of positive, zero, and negative constant curvature.[3]

As early as 1942 Milne noticed an important *conceptual* feature of these Newtonian cosmological models.[4] The models are constructed by assuming that an observer at the center of the expanding sphere of matter is in an inertial frame. Since he is central he feels no net gravitational field from the matter surrounding him. Any other observer who is "co-moving," i.e., is fixed to a particle of the expanding matter sphere, will be accelerated with respect to the central observer, and hence not an inertial observer. Furthermore, he will experience a net gravitational field induced at his location by the remaining matter of the universe. This seems to make the central observer highly privileged in this model and to violate any form of a "cosmological principle" which makes all co-moving observers "equivalent."

But, Milne argued, this alleged inequivalence fails to hold in the Milne-McCrea models. An examination of the models shows us this: Start with *any* co-moving observer. If he assumes himself inertial, and postulates a gravitational field which assigns to each other point the gravitational field generated by the matter in the sphere marked out by the radius between this second point and our initial observer, then he will obtain as his cosmological solution the same description of the universal history, relative to his observational stance, as would *any other* co-moving observer who did the

1. For references, see O. Heckmann, *Theories der Kosmologie: Berichtiger Nachdruck* (Berlin: Springer-Verlag, 1968), 1, and O. Heckmann and E. Schücking, "Newtonsche und Einsteinsche Kosmologie," in S. Flügge, ed., *Encyclopedia of Physics,* vol. 53 (Berlin: Springer-Verlag, 1959), 491 n. 3.

2. E. Milne, "A Newtonian Expanding Universe," *Quarterly Journal of Mathematics, Oxford Series* 5 (1934): 64–72.

3. W. McCrea and E. Milne, "Newtonian Universes and the Curvature of Space," *Quarterly Journal of Mathematics, Oxford Series* 5 (1934): 73–80.

4. E. Milne, "Absolute Acceleration," *Nature* 150 (1942): 489.

same thing. At least this will hold so long as one has "gone to the limit" where the radius of the "sphere" of expanding matter has been allowed to "go to infinity."

This is an important conceptual point. What it tells us is that, in certain cases at least, one can "trade off" noninertiality and gravitational fields. Although, from the strict Newtonian point of view, only one co-moving observer is, in general, Newtonian, since, in general, the co-moving observers are relatively accelerated, each co-moving observer can treat the universe "as if" he were Newtonian. To do this he must, from the Newtonian point of view, at the same time *mis*represent the real gravitational field, among other things, by acting as though the field at his location were null when it really is not.

In 1937 Synge had already made the point that within the Newtonian framework such a trade-off between noninertiality and the postulated field existed.[5] For consider a world in which a uniform gravitational field of force exists, i.e., a world with a gravitational potential whose gradient is in the same direction and of the same magnitude everywhere. A reference frame in "free fall" in this field will be accelerated relative to the Newtonian inertial frames. But suppose an observer attached to such a frame decides to consider himself inertial and tries to describe the world according to the usual, not fully covariant, equations of Newtonian mechanics. So long as he attributes to the world a null gravitational field everywhere, instead of the non-null, constant field which "really" exists, he will get the right results. This point was already noticed by Maxwell as early as 1877.[6]

Further development of the Milne-McCrea approach took place in the 1950s in McCrea's response to remarks of Bondi and Layzer.[7] In his responses McCrea argued that the Milne-McCrea approach could not be faulted for assuming that mutually accelerated observers were both really inertial, for, he said, the theory really took only one co-moving observer to be inertial and only

5. J. Synge, "On the Concept of Gravitational Force and Gauss' Theorem in General Relativity," *Proceedings of the Edinburgh Philosophical Society, 2nd series* 5 (1937): 93–102.

6. J. Maxwell, *Matter and Motion* (New York: Dover, 1954), 85.

7. H. Bondi, *Cosmology* (Cambridge: Cambridge University Press, 1960), 78; and D. Layzer, "On the Significance of Newtonian Cosmology," *Astronomical Journal* 60 (1955): 268–269.

took the Newtonian equations to apply in his inertial frame.[8] The "equivalence" of other co-moving observers, allowing them to use Newtonian equations if they modified their gravitational field attribution from that of the true inertial observer, was a *consequence* of the assumed Newtonianism as expressed in the preferred frame, not an *assumption* that these observers were Newtonian.

McCrea did allow, however, that the device of passing from world models in which the matter formed a sphere of large, indeterminate but finite radius to that in which it filled Euclidean space was illegitimate from the strict Newtonian point of view. For the usual Newtonian approach, which proceeds by solving Poisson's equation for the gravitational potential relative to suitable boundary conditions, breaks down in the "infinite distribution of matter" limit. Yet, he argued, the Milne-McCrea model could still be called Newtonian in a limited sense. It is not true that one could, in this model of a matter-filled infinite universe, find a potential from which the forces could be derived where the potential is uniquely determined by the matter distribution, as one could in the "island universe" case by solving the Poisson equation with the boundary condition of the potential going to zero at infinity. Yet in the Milne-McCrea type models it is still the case that the matter obeys Newton's laws relative to an inertial frame and that the forces can be deduced from *some* potential, not uniquely determined by the matter distribution, which obeys Poisson's equation.[9]

In any case, he argued, we could always drop the "limit as the matter extends to infinity" case and deal with a sphere of matter of large, indeterminate but finite radius. In such a Newtonian world the Milne-McCrea picture of a world which obeyed the "cosmological principle" of all co-moving observers being suitably situated for "equivalently" describing the world would still hold for all observers far from the boundary of the matter sphere. For insofar as he limited his description to the behavior of observed matter in the interior of the sphere, and made no observations on, for example, his situation relative to the sphere boundary, each observer could, again, use Newtonian mechanics in a suitable (nonrotating) refer-

---

8. W. McCrea, "On Newtonian Frames of Reference," *Mathematical Gazette* 39 (1955): 287–291; "On the Significance of Newtonian Cosmology," *Astronomical Journal* 60 (1955): 271–274; and "Newtonian Cosmology," *Nature* 175 (1955): 466.
9. McCrea, "On the Significance of Newtonian Cosmology," 273.

ence frame attached to his own co-moving point of matter and, with a suitable choice of gravitational field (essentially, making it zero at his own location and letting the force felt at another location be that due to the matter in the sphere marked out by the radius from the observer to the new location), give a correct description of the behavior of all observed matter.

Building on these developments, Heckmann and Schücking provided an approach to Newtonian cosmology which at the same time gave a new conceptual approach to Newtonian mechanics and theory of gravitation in general.[10]

Consider the notion of the Newtonian inertial frames. How can we determine which frames there are? In the "island universe" case we might try to do this by using freely moving particles "at infinity," or, rather, far removed from gravitating matter. But the Maxwell-Synge argument shows us that even in this case we could not in any experimental way be sure that the particles were *really* free, and not subject to a uniform gravitational field. (The exact proportionality of gravitational force to inertial mass and the universality of the gravitational force are, of course, crucial here.) So we could not be sure that we had picked out with our "free" particles the inertial systems, and not a frame in uniform acceleration with respect to the inertial frames.

In the case of the matter-filled universes so crucial to cosmology, finding even nominally free particles "far removed from all matter" becomes an impossibility.

So why not drop the notion of "the inertial frames" altogether? And with it drop the notion of "the gravitational field." For the former substitute the notion of the *local inertial frames*. To find his local inertial frame an observer finds a frame not in rotation, as evidenced by the absence of centrifugal and Coriolis forces, which "falls free" in his local gravitational field. We assume that Newton's laws in their usual noncovariant form hold in each of these local frames.

But we must drop the notion of "the gravitational field" as well. Each local inertial observer sets the gravitational field at his loca-

10. O. Heckmann and E. Schücking, "Bemerkungen zur Newtonschen Kosmologie, I," *Zeitschrift fur Astrophysik* 38 (1955): 95–109; "Bemerkungen zur Newtonschen Kosmologie, II" *Zeitschrift fur Astrophysik* 40 (1966): 81–92; "Newtonsche und Einsteinsche Kosmologie," sec. II.

tion at zero. But, in general, as is the case in the general Milne-McCrea models with the co-moving observers, the local inertial observers will be accelerated with respect to each other. Each observer blames the acceleration of the others on the presence of a nonzero gravitational field at these distant locations. Each local inertial observer can give a correct account of the motion of matter by using the usual Newtonian laws in his local inertial frame. But this can be true for all of them only if each postulates a gravitational field whose values at a given point will vary from observer to observer.

Using this "local Newtonian" approach Heckmann and Schücking go on to show how one can drop the usual boundary conditions necessary for solving the Poisson equation in the usual Newtonian approach, and replace them with conditions invariant, as the usual are not, among the local inertial observers. They show how these new constraints can be applied even in the case of matter-filled universes, and by so applying them construct not only the previously known Newtonian analogues for the Friedmann solutions of general relativistic cosmology, but analogues for the Gödel solution in which the averaged out mass of the universe is in absolute rotation as well. Indeed there is a principal virtue of this "localized Newtonianism," in which the notion of "the inertial frames" and the "gravitational field" is exchanged for a field of local inertial frames, and a notion of gravitational field is determinate only relative to a particular local inertial frame. Since "localized Newtonianism" is formulated in a manner highly reminiscent of general relativity (invoking only the *local* notion of inertial frame and utilizing at its core the mechanical aspects of the principle of equivalence), a convenient comparison of related Newtonian and general relativistic cosmological models is easily made in its terms.

III

The "local Newtonianism" of the last section, with its eschewal of the notion of global inertial frames and its relativization of the field to the reference frame chosen, and with its close analogy to the well-known features of general relativity, cries out for a four-dimensional spacetime formulation. Work in this direction begins with Frank's four-dimensional formulation of Newtonian and special relativistic mechanics without gravity, and is generalized by

Cartan and Friedrichs as early as the twenties to encompass Newtonian gravitation.[11] Beginning in the sixties this work has been reformulated in the coordinate-free notation of contemporary differential geometry.[12]

Such a formulation will, first of all, posit spacetime to be a four-dimensional differential manifold, homeomorphic to E^4, with a symmetric affine connection. The latter defines the geodesics of the spacetime. The spacetime will have a form t_a, sufficient to guarantee an "absolute time," and it will be posited that the covariant derivative of this form is zero. This latter condition implies that the form is, locally, the differential of a function ("local time") which has the character of an affine parameter along the time-like geodesics, and time measurement will then be unique up to a linear transformation, i.e., up to a choice of a zero point of time and a unit of measure. The spacetime is also endowed with a contravariant tensor, h^{ab}, of signature $+++0$ whose covariant derivative also is zero and which is such that $h^{ab}t_b = 0$. This tensor serves as the metric tensor of the three-spaces into which the spacetime is "stratified" by taking as spacelike slices subspaces of the spacetime with constant values of the time measure.

We need additional mathematical constraints as well, however. In the Trautman version, these are two equations connecting the temporal form and the spatial metric tensor respectively with the Riemann curvature tensor of the spacetime and are needed to guarantee that we can get back out of the spacetime the desired local Newtonian theory of mechanics and gravitation. Essentially, the conditions are sufficiently strong to guarantee that we can decompose the affine connection of the spacetime into two parts: one part the integrable affine connection characteristic of a flat spacetime (i.e., one whose Riemann tensor is identically null), and the other dependent upon the time form and the *gradient of a scalar*

11. For references, see P. Havas, "Four-Dimensional Formulations of Newtonian Mechanics and Their Relation to the Special and the General Theory of Relativity," *Reviews of Modern Physics* 36 (1964): 938–965, esp. p. 939 nn. 14, 15.

12. A. Trautman, "Comparison of Newtonian and Relativistic Theories of Space-Time," in B. Hoffman, ed., *Perspectives in Geometry and Relativity* (Bloomington: Indiana University Press, 1966), and "Sur la Théorie Newtonienne de la Gravitation," *Comptes Rendus* A257 (1963): 617–620. See also C. Misner, K. Thorne, and J. Wheeler, *Gravitation* (San Francisco: Freeman, 1973), chap. 12.

function only. We shall discuss this decomposition presently. The axioms also guarantee that the metric geometry of the simultaneity slices will be Euclidean. There are "interpretive axioms" (or, according to some, *definitions*) which tell us that ideal clocks measure the affine parameter of time, that ideal rods measure spatial distance as determined by the spatial metric tensor, and that free particles (i.e., free of all but gravitational "force," now absorbed, as in general relativity, into the geometry) travel timelike geodesics. Notice that the spacetime we have described is *not* a Riemannian or pseudo-Riemannian spacetime in any sense, for while there is an invariant temporal metric and a spatial metric on the simultaneity subspaces, there is no invariant spacetime separation between arbitrary events of the spacetime, i.e., there is no spacetime metric at all. Finally, there is the surrogate to Poisson's equation which expresses, now in terms of the Ricci tensor, the relation of the local "gravitational field" to the local density of matter.

The decomposition of the affine connection, noted above, has an obvious physical interpretation. The flat part of the affinity characterizes a flat neo-Newtonian spacetime. This spacetime is not that of Newton with its "absolute space," for sameness of place of two nonsimultaneous events is not defined in it. To get that we would have to add a "rigging" to the spacetime, i.e., an assignment of timelike world-lines which would tell us which nonsimultaneous events were happening at the same spatial place. But it is a spacetime in which the global inertial frames are well defined. They are fixed by the straight-line geodesics of the now flat spacetime. The scalar field which then, in conjunction with the temporal form, fixes the remainder of the affinity, can be interpreted as the gravitational potential. So the decomposition gives us the picture of free particles traveling straight-line geodesics in neo-Newtonian spacetime with its global inertial frames, where by 'free' we now mean free of gravity as well as of other forces. And particles acted upon by gravity can be taken to be "forced" out of their geodesic motion by the force measure as the gradient of the gravitational potential.

But it is crucial to realize that while the axioms of local Newtonian spacetime do guarantee that such a decomposition of the affine connection is possible, they most certainly do not show that such a decomposition is unique, and indeed it is not. As we noted in section II we have the option of extending different, and mutually relatively accelerated, local inertial frames into the global inertial frame of our neo-Newtonian spacetime, if, at the same time, we are willing to pick a gravitational potential whose spatial form is appropriate to our choice of inertial frames. When two local inertial frames are in relative acceleration, of course, the appropiate scalar potential functions will not have the same spatial distribution.

As before, in the cosmological theories with matter distributed throughout the simultaneity spaces, we have no "natural" way of picking the appropriate distribution, unless, out of pride of place perhaps, we pick the local inertial frame where we are as "the" inertial frame. In the case of the island universe we can, if we wish, pick the decomposition which takes the local inertial frame far removed from any matter as "the" inertial frame. But, as before, we need not do this, for even in this case we can, should we so choose, describe the world as imbued throughout with a uniform gravitational field, so that the local inertial frame "at infinity" does not pick out the "real" flat affinity, or the real global inertial frame. In any case, it is important to note that what we can empirically determine is the local inertial frame. The affine connection of the spacetime is not "arbitrary," whereas its decomposition into flat affine connection and gravitational field is, at least in the sense that no local, mechanical observations determine which decomposition to choose.

IV

It is one thing to paraphrase results of mathematical physics but another to draw philosophical consequences from them, and it is to this latter problem that I now turn. My focus will be obtained by reflecting upon, and commenting on, some recent claims of Earman and Friedman (hereafter referred to as E&F) in their recent enlightening examination of the status of Newton's First Law of motion in the light of contemporary spacetime physics.[13]

E&F argue against the conventionality of Newton's Law of Inertia and related principles; and they argue for the thesis that the experimental data which would support Newtonian physics would also support a view of spacetime as a "substantival" entity (as would, they claim, the data grounding belief in general relativity). They also argue against the possibility of constructing what I shall call Machian relationist substitutes for the Newtonian theories and against the thesis that the Law of Inertia is in any sense idiosyn-

13. J. Earman and M. Friedman, "The Meaning and Status of Newton's Law of Inertia," *Philosophy of Science* 40 (1973): 329–359.

cratic among the other theoretical laws of nature. Finally, they argue that there are crucial differences between the "conventionalistic" aspects of the Law of Inertia in Newtonian physics and the laws which take its place in general relativity.

I most certainly agree with them about the impossibility of "Machianizing" Newtonian physics, but I believe that they are wrong in thinking that this supports a general antirelationist thesis. And, in general, I think that their arguments for substantivalism rest upon hidden metaphysical assumptions which should be displayed if we are to understand the true relationship between Newtonian physics and spacetime substantivalism. While I agree with them that the inertial principle is of a piece with other theoretical laws of science, I do not believe that they do full justice to conventionalist claims, nor to just what is required to resolve the conventionalist issue in either direction. Finally, while I agree that there are crucial differences between general relativistic spacetimes and those of Newtonian mechanics, I believe that there are far greater similarities than they admit with respect to the issues both of conventionality and of substantivalism versus relationism.

Let me begin by outlining how someone might make the transition from one Newtonian spacetime to another; in particular, from Newton's spacetime to flat neo-Newtonian spacetime to local Newtonian spacetime. To start I am going to make the bold assumption that the only observable quantities in our theory will be the *relative* positions, velocities, accelerations, etc., of material objects. And I will eschew any mention of electromagnetic phenomena. Later attention will be paid to both these points.

Noticing the existence of forces (as displayed in, for example, sloshing water, bulging equators, etc.) not accountable for in terms of the relation of the affected object to other material objects, and the apparent dependence of these forces upon accelerations of some kind, Newton posits substantival space and blames the forces on acceleration relative to this entity.

But since the forces depend only upon "absolute" accelerations, and not on "absolute" velocities, we notice that we could take as substantival space an object at rest in any of the inertial frames. To pick one such frame as the rest frame of substantival space invokes a "convention," in at least one sense of that word: to wit, an arbitrary choice.

Equipped with the notion of a *spacetime,* originally designed for *special* relativity, we discover that we can strip our Newtonian spactime theory of some of its structure, thereby obviating the need to make a conventional choice of rest frame for substantival space. In this neo-Newtonian spacetime, equipped with absolute time, three-dimensional Euclidean spaces and an integrable affine connection, but devoid of the "rigging" which fixed substantival space, we can still account for inertial forces as the result of motion deviating from timelike geodesics, yet we need no longer conventionally choose which inertial frame is the rest frame of a (now no longer postulated) substantival space.

But next we notice the Maxwell-Synge problem. Can we really determine which are the inertial frames, without making, again, a "conventional" choice? Not if we are willing to countenance universal and uniform fields of force. Perhaps this would not bother us so much, except that we know of a "real" force whose behavior is too close to our "imaginary" force for comfort. For gravity is, at least, universal, if it still remains the case that no "real" material sources of gravitation can generate a uniform field of force. The existence of gravity turns the philosopher's gimmick of a "universal force" into a real problem *physics* must accommodate.

And cosmological considerations require us to take the Maxwell-Synge observation even more seriously. In the "island universe" case we might take the "conventional choice" of setting the gravitational field far from matter equal to zero, and thus "conventionally" pick out the inertial frames. Such a "choice" is so natural that we might not even notice that we were making it. But in the case of matter-filled universes this choice is not open to us. Now the only "natural convention" is to pick our local inertial frame as globally extendible to the "real" overall inertial frames. But such a choice immediately strikes us as being at variance with our desire to treat all (co-moving) observers equivalently.

Then, equipped with the insights of *general* relativity, we see that once again the problem of conventional choice can be obviated. We do this by "geometrizing" the gravitational field into the spacetime structure itself, leaving no choice of field to be made independently of picking the spacetime structure. And by allowing for curved spacetimes, we go over to a spacetime which has only local inertial frames. Since there is no global inertial frame to be discovered, we need not choose it. We need only empirically discover the affine connection of the curved spacetime at each point. Of course, having mapped out the spacetime in this way, we could now "decompose" the affine connection into flat connection and gradient of scalar field, but such a decomposition no longer appears as making a decision about which frames really are the global inertial frames, but, instead, rather looks more like a mere "choice of coordinate system."

There is an interesting difference between the change from Newtonian to neo-Newtonian spacetime and that from neo-Newtonian to local Newtonian. In the first case we need only drop one spacetime for the other, arguing that the former had superfluous elements all along. In the latter case we cannot just modify our spacetime theory, but must modify our gravitational theory as well. It is as though we "distributed" the conventional element in our theory over two apparently distinct portions of physics in the second case and had to *discover* this fact, whereas in the first the conventionality could be located in what was all along recognized as a single segment of our theoretical structure.

But suppose someone, believing one of the earlier theories about spacetime, and presented with the later of the pair, refuses to accept the conclusion that he should relinquish his earlier theory for the newer. Suppose Newton, for example, after learning about neo-Newtonianism, still refuses to acknowledge the nonexistence of substantival space (or the nonexistence, in the world, of the "rigging" of neo-Newtonian spacetime as representing a real element of nature). Again imagine a believer in flat neo-Newtonian spacetime who, when presented with local Newtonianism, still insists upon the existence of the global inertial frames. How could such a reactionary argue? Let us, for specificity, stick to the man who holds to neo-Newtonianism as against local Newtonianism.

Faced with the island universe case he is likely to argue that the "real" inertial frames are those generated by extending the local inertial frame of observers remote from all matter. This, he will argue, is the simplest possible theoretical choice. Faced with the cosmologies of filled universes, however, his problems are greater, for here he seems to have no ground for extending any particular local inertial frame to "the" global frame, lest it be parochial egocentricity. But, he can argue, there must be some global frame. So the cosmological principle is wrong. Only one co-moving observer is privileged in having his local inertial frame coincide with the "real" inertial frames. But is this really as horrendous an assumption, he will argue, as the thoroughly implausible assumption, say, that the earth is the physical center of the universe, the kind of assumption the cosmological principle was really designed to rule out? In any case, he continues, an "observational cosmological principle" will still hold in the sense that any co-moving observer who takes his local inertial frame to be the true inertial frame, and who appropriately calculates the gravitational field, will still get a correct description of the universe should he apply the Newtonian equations in his local frame. So we cannot *tell* which local frame is really inertial.

Now the response to this position, with its invitation to conventionalists, is fairly obvious. In the filled universe case we are likely to argue that the choice of which local frame is truly inertial is totally undetermined by any possible observational facts. So really, we are likely to continue, there is no choice to be made at all. Any such "choice" is merely a trivial rewording of one and the same theory. This is, of course, the familiar response to allegations of conventionality of Eddington, Reichenbach, and Grünbaum. And is the situation really so different in the island universe case? Here also there exist, from the point of view of the exponent of neo-Newtonianism, an infinite number of "alternative" theories all "saving the phenomena." Each theory postulates a set of inertial frames and a uniform force field throughout the universe — but a different force field, and a different set of inertial frames (these having the appropriate relative accelerations, corresponding to the differences of the fields), for each different theory in question. These are all the Maxwell-Synge "alternatives." Once again, the claim will be that this empirical underdetermination reflects only trivial semantic conventionality. Indeed, this is just the line E&F take with regard to a related proposal (in their "corrected theory of Ellis").[14]

So, one continues, isn't the move to local Newtonianism the reasonable thing to do? In the light of this theory of spacetime the merely semantical alternatives which previously masqueraded as real alternatives show their true nature. There are many ways to decompose the affine connection into a flat, integrable part and a remainder which depends upon a gradient of a scalar field. No such decomposition represents a different reality. They are all just different ways of expressing the same thing, the affine connection,

14. Earman and Friedman, "Meaning and Status," 347.

which is the linguistic representative of the full structure really in the spacetime.

But we must be cautious here. For what about the shift from any one of the neo-Newtonian spacetimes (or, rather, the whole class of them) to local Newtonian spacetime? Any particular neo-Newtonian spacetime will give the same predictions as the local Newtonian theory. After all, the former can be derived from the latter simply by making one of the permissible decompositions of the affine connection. So isn't the choice between this (or any) neo-Newtonian theory and local Newtonianism "merely a trivial semantic matter"? In any case, isn't someone (1) who is prepared to fault the neo-Newtonian theory on the grounds of all the "trivial semantic alternatives to it masquerading as real theoretical alternatives," and (2) who is still insistent that the shift from neo-Newtonianism to local Newtonianism is a "real theoretical shift," obligated to provide us with some criteria of what counts as a trivial semantic rewording and what as a genuine theoretical shift when, in both cases, *no differential empirical predictions are involved?*[15]

Now the move from neo-Newtonianism to local Newtonianism is usually going to be based upon some kind of invocation of Ockham's Razor. The postulated global inertial frames of neo-Newtonianism add no more explanatory power to the theory than we already possess when we have a spacetime equipped with an affine connection, i.e., with local inertial frames. So why not adopt the theory which accounts for the data with the fewest theoretical entities postulated?

But now it is only fair to raise the question: Have we gone far enough? We disposed of substantival space in favor of the inertial frames in going from Newtonian to neo-Newtonian spacetime. And we disposed, simultaneously, of gravitational forces and global inertial frames in going from neo-Newtonian to local Newtonian spacetime. But we are still postulating a spacetime structure which has a causal influence on the material objects immersed in it, as E&F so clearly point out. Can we go further and dispose of spacetime as an entity in its own right altogether? In particular, can we dispense with it in favor of a purely relational theory of spacetime, in which the only existing objects are the material objects,

---

15. For more on this, see L. Sklar, *Space, Time, and Spacetime* (Berkeley: University of California Press, 1974), 113–146.

which retain, of course, their spatiotemporal relations to one another, but which now have no relations to "spacetime itself" (for there is no spacetime itself)?

Now a motivation for getting rid of global inertial frames was that we could, by postulating uniform universal forces, choose any frame we liked as "really inertial." Better to do without such a postulation altogether than to infect one's physics with such arbitrariness, we argued. But does the same problem arise with the affine connection of local Newtonianism as arose with the global inertial frames of neo-Newtonianism?

Indeed it does—but with a difference. The Maxwell-Synge problem already showed a difficulty with neo-Newtonianism. Notice that it worked by considering the possibility of postulating *uniform* universal forces, forces unlike anything normally spoken of in physics. The difficulties neo-Newtonianism faced in the cosmological models of filled universes showed us that gravity, the ordinary gravity we normally postulated in Newtonian physics, caused difficulties for neo-Newtonianism as well. This made us take the Maxwell-Synge "forces" much more seriously than we would have otherwise. Essentially, it converted them from a philosopher's gimmick ("universal forces") to something closely related to a genuine physical problem — how to treat gravity in Newtonian cosmology.

In playing the same game with local Newtonianism we have only the philosopher's gimmick with which to work, and nothing like the real physical problem facing neo-Newtonianism. The gimmick exists nonetheless. The local Newtonian theory postulates that free particles (free of all but gravitational forces, of course) travel timelike geodesics. So to find the timelike geodesics of the spacetime use free particles. But how do we know which particles are really free? Not by seeing which travel timelike geodesics, surely. So we can posit any set of events we like as lying along a timelike geodesic, so long as we postulate sufficient fields of "forces" to deviate what we ordinarily took to be free particles to their observed paths and away from our postulated timelike geodesics. The point is philosophically obvious, and the construction of the appropriate "fields" both trivial and unnecessary.

The problem is the standard one. So long as we introduce into our theory any elements not "directly accessible to observation" but to which epistemic access can be obtained only through their relation to some observable entities, we can always postulate some new relation of unobservable to observable by putting enough new theoretical apparatus into the theory to introduce new "dualities."

Now we have been assuming all along that the observables of our theory have been material objects and their spatiotemporal relations. So could we now move from local Newtonianism to a purely relational theory, eschewing the postulation of an "unobservable" spacetime itself altogether?

One way this *cannot* be done, at least in the case of Newtonian theory, as E&F among others have so clearly pointed out, is by moving to Machian relationism. Within the context of Newtonian theory, and this is what we are trying to understand, not what the formulation of some theory empirically incompatible with Newtonian mechanics might look like, Newton's original arguments about weights on the end of a string in an otherwise empty universe would suffice to refute any Machian reformulation which attempted to account for inertial forces by the relation of material objects to one another rather than by their relations to the structure of spacetime itself.

But Machian relationism is not the only kind of relationism, and there is another variety, trivially formulated and explained, which will do the trick. We simply do not explain the inertial forces. Some objects feel them and others do not. What we do explain is the difference in felt inertial force in terms of the relative motion of objects. An object rotating with respect to a frame at rest in an object feeling no centrifugal and Coriolis forces experiences such forces. Our theory tells us the correlation between the amount of inertial force felt and the amount of relative rotation. It does not tell us why one object feels the forces and the other does not. It simply does not explain this fact but takes it as a fundamental fact of nature. For, it argues, the only explanation the other theories gave was in terms of some motion (or absence thereof) of the objects relative to an otherwise totally unobservable spacetime, and, being devoid of predictive value, that explanation is no explanation at all.[16]

So we can go from Newtonian to neo-Newtonian spacetime, from neo-Newtonian spacetime plus gravity to local Newtonian spacetime, and from local Newtonian spacetime to a theory which does not posit spacetime as an entity in its own right at all. The only

16. For some more material on relationism, see L. Sklar, "Absolute Space and the Metaphysics of Theories," *Nous* 4 (1972): 289–309, and *Space, Time, and Spacetime*, chap. 3, esp. sec. F.

way I see of excluding this kind of relationism would be to invoke a *metaphysical* assumption which would demand that inertial forces be explained by the motion of the object suffering them relative to *some* reference object.

It might be useful to compare this last transition to the earlier two. Each transition reduces our ontological commitment. The first drops substantival space, the second global inertial frames and the gravitational field, and the third spacetime as an entity altogether. While the third is, perhaps, the philosophically most dramatic transition, it is certainly the least interesting mathematically and physically. While the first transition forces us to go from space and time to spacetime in the mathematical apparatus needed to represent our theory, and the second to go from flat to curved spacetime, the third transition introduces no new interesting mathematical formulation at all. Physically the second transition is the most interesting, for unlike the first and third which work almost entirely by "philosophical insight" (you can do without absolute space and without spacetime altogether and get the same results, empirically, in your theory), the second by "geometrizing" gravity into the spacetime structure makes the biggest difference in (at least in the appearance of) our physical theory. And, although we can do Newtonian cosmology without moving to local Newtonian spacetime, cosmology is certainly most elegantly done in the framework of local Newtonianism.

At this point it is useful to move beyond the consideration of spacetime theories to look at what we have been discussing from a more abstract and purely philosophical point of view.

The general situation is this: We have a theory which serves as an explanatory account of some observable features of the world. The theory posits the existence of entities not themselves observable. This allows for the possibility of alternative accounts, also positing unobservable entities, which explain the phenomena equally well. Hence, it is soon alleged, there is an element of conventionality in our theory of the world. Soon someone proposes that the alternative accounts really are not alternatives at all, but trivial semantic variants of one another, which appear distinct only because of equivocation on the meaning of crucial terms. Finally, it is proposed that we move to a new theoretical account which, by ontological retrenchment, i.e., by positing a sparser unobservable ontology, removes some of the opportunity for underdetermination and hence for (apparent) conventionality.

The basic problem with this approach is always the same: Where do you stop? If one rejects absolute space, global inertial frames, and even, as I suggest can be done, spacetime as an entity altogether, why not make the final retreat to phenomenalism or, worse, solipsism of the present moment? What we have seen above is one more example of the slippery slope philosophers enter onto when once they begin an epistemic critique of any realistic theory. What seems to be the case is that "dualities," "underdetermination," and, hence, "conventionality" are the price of scientific realism. We avoid them altogether only at the price of eschewing a realistic approach to science.

Now maybe at some point we want to block the regress by arguing that although two theoretical accounts lead to no alternative observational consequences, it is theoretically reasonable to adopt one rather than the other, a view which presupposes, of course, that they are genuinely distinct theories. I sympathize with this view, but what I want to suggest here is that we are at least obligated to be consistent. It seems unfair to claim in one case that some alleged theoretical alternatives between which we must "conventionally" choose are mere semantic variants, and then dispose of the problem by ontological retrenchment, while in the next case we argue that there simply is a theoretical choice which must be made between two genuinely distinct theories which may be made on "theoretical" grounds, despite the fact that the theoretical alternatives are just as observationally equivalent in this latter case as in the former. It is unfair, that is, unless one has found some, as yet overlooked, crucial distinction between the cases. Yet I suggest it is exactly this kind of philosophical inconsistency which E&F get into when they first reject what they call the "corrected Ellis theory" as a mere semantical variant of Newton's and not a genuine theoretical alternative, yet opt for local Newtonianism over neo-Newtonianism (as I think they would do) and for local Newtonianism over my variant of relationism as they must do if they are to maintain their claim that their paper "provides support for a realist interpretation of space-time theories."[17]

---

17. Earman and Friedman, "Meaning and Status," 329.

I have argued that the problems associated with philosophical doctrines of conventionality are more pervasive and intractable than might at first appear, and I have suggested that in general it might be hard to reconcile anticonventionalism with physical realism as E&F try to do. There are other arguments against conventionalist theses which deserve a little of our attention though, not least because E&F briefly mention them. Let me look at three such arguments.

(1) "The observational/non-observational distinction [is] vague . . . [and] it is constantly shifting."[18] Indeed it is, at least as long as we do not go to the natural limit noted above and take as observable only sense-data or the like. We can, of course, always try to undercut allegations of conventionality by counting as at least potentially observable every entity posited by our theory. I doubt that we would be persuaded to do this in many cases: would anything convince us, for example, that we really should view Heisenberg quantum mechanics and Schrödinger quantum mechanics as genuine alternatives possibly discriminable by "direct observation" of the wave-function?

But in any case, once again, at least we can ask for consistency here. E&F themselves speak of the "corrected Ellis theory" as a trivial semantic variant of Newton's. But how could they possibly think that, unless they themselves are quite convinced that the "real inertial frames" of spacetime are completely beyond any "direct" observational test? It seems unfair on the one hand to object to conventionalist moves by claiming that the observational/nonobservational distinction is vague and shifting, and yet in other places attempt the standard refutation of conventionalism which goes by arguing that the "alternatives" are merely semantic alternatives, and which presupposes the observational/nonobservational distinction just as much as conventionalism does.

Again consider the arguments which lead us from neo-Newtonianism to local Newtonianism. E&F would, I think, speak of this as a change of "theoretical commitment." But just what do the usual arguments look like in *actual scientific practice?* I think that examination will show that the arguments look like this: Anything explicable by neo-Newtonianism is explicable by local Newtonianism. But the latter makes less of a theoretical commitment to enti-

18. Ibid., 348.

ties unobserved than the former, requiring only an affine connection and not requiring global inertial frames or a unique gravitational field. Therefore the latter is preferable to the former.

It is clear that this argument rests upon presuppositions about just what is and what is not observable. Indeed, it presupposes that material objects, their mutual spatiotemporal relations, and the effects generated in them by inertial forces are observable. And it assumes that spacetime itself and the gravitational field itself are not.

What I am arguing is that it is one thing for philosophers reluctant to submit to the rigors of the conventionalist argument to deny the plausibility of a definite and fixed observational/nonobservational distinction. But what they must do is explain the pervasive role the presupposition of just such a distinction plays not just in positivist philosophy of science but in science itself. And it is unfair to deny the distinction when it hurts but then to argue "from science" when the very science one is using is frequently implicitly or explicitly based on the presupposition of just such a hard and fast distinction.

(2) "In enlarged contexts these commitments can give rise to different observational consequences."[19] Here they have in mind something like the invocation of electromagnetic phenomena to distinguish among spacetimes which are mechanically observationally equivalent. But I do not think that any determined conventionalist or antisubstantivalist will be taken in by their argument.

In the nineteenth century, for example, it was thought that optical experiments might tell us which inertial frame was the rest frame of substantival space, for the velocity of light in Newtonian theory can only be isotropic in one inertial frame, and not in those moving uniformly with respect to it. But, of course, even from an aether theory point of view, the best a Michelson-Morley experiment can tell us is what the rest frame of the aether is. But why identify this with substantival space? We have already seen that the theoretical move initiated by Newton and continued throughout neo-Newtonian and local Newtonian theories to account for dynamic effects (inertial forces) by the relation of matter to substantival space or spacetime is not in any way a necessary feature of

19. Ibid.

physics. So dynamical effects cannot even assure us that spacetime exists as an entity over and above the matter in it. And surely, a conventionalist would argue, we could account for the positive results of the Michelson-Morley experiment taking any of the spacetime theories we have considered that we like. Only if we make the gratuitous and empirically totally unmotivated, i.e., conventional, decision to identify the aether frame with the frame of substantival space can electromagnetism make detection of the rest frame of substantival space possible and so rationalize the choice of Newtonian as opposed to, say, neo-Newtonian spacetime.

The example E&F give is somewhat different.[20] They write Maxwell's equations in covariant form and remark that these equations have different consequences depending on whether the covariant derivatives involved are determined by the flat affinity of a neo-Newtonian spacetime or the nonintegrable affinity of a local Newtonian spacetime. But so what? Suppose the equations give the right results only with respect to a flat affinity. Does this preclude us from believing in a curved local Newtonian spacetime, but nonetheless introducing equations for electromagnetism which require covariant differentiation with respect to a flat affinity? Why should it? Once again, I submit, electromagnetism, like dynamics, can tell you something about the structure of spacetime only if you let it. Only if you build in enough theoretical (metaphysical? conventional?) assumptions so that the observational results are interpreted in one way, rather than in the other equally possible ways they would be read if you put in different presuppositions.

(3) "Forces have causes as well as effects, and the causes are often due to sources which are as accessible as their force effects."[21] Here the argument seems to be something like this: Consider an alleged pair of conventional alternative accounts, within neo-Newtonian spacetime for example, about just which frames are inertial. B claims as an inertial frame one in relative acceleration to a frame designated by A as inertial. When it is pointed out to B that material objects under the influence of no other material objects accelerate uniformly with respect to his "inertial frame," he explains this by referring to a Maxwell-Synge uniform universal force of gravity.

20. Earman and Friedman, "Meaning and Status," 349.
21. Ibid., 346.

But gravity cannot exist like that, E&F seem to reply, for gravitational fields have their sources in really material objects and are never uniform. Hence B is wrong.

But of course this will have no impact whatever on the conventionalist. He knows that sometimes if we are to preserve the right observational consequences of a theory while at the same time making changes in its theoretical structure, we must make changes elsewhere in the theory as well. Of course B will have to change his theory of gravity (allowing uniform source-free gravitational fields as well as the usual ones) if he wishes to make a nonstandard choice of inertial frame. But that is a familiar feature of the conventionalist argument — not a refutation of it.

V

Let us now turn our attention from the various spacetime approaches to Newtonian theory to general relativity. It is reasonable to suppose that if in each version of Newtonian theory conventionalistic problems arise, and if, in each version of Newtonian theory, antisubstantivalist arguments can still be reasonably maintained, then the same conventionalistic and relationistic arguments might apply to the spacetime of general relativity as well. Indeed, I believe that they do, and in a way which leaves them *philosophically* almost unchanged. But this has been disputed by E&F, so we should look at their arguments with some care.

It is certainly true that the spacetime of general relativity is *scientifically* different from those of the Newtonian spacetimes in interesting ways. Like local Newtonian spacetime, and unlike Newtonian and neo-Newtonian spacetime, it is a curved spacetime. Unlike local Newtonian spacetime, however, the spacetime of general relativity is a pseudo-Riemannian spacetime and as such has a four-dimensional (interval) metric. Whereas local Newtonian spacetime looks locally like neo-Newtonian spacetime with its spaces and absolute times, general relativistic spacetime looks, locally, like Minkowski spacetime, with its familiar interval metric.

Now E&F argue as follows: In any spacetime if, indeed, it is a metric spacetime, it seems reasonable to insist that the affine connection be compatible with the metric. What this means is that the covariant derivative of the metric tensor field be null with respect to the affine connection. Or, geometrically, that "parallel" transport of vectors does not change their length. But general relativistic spacetime has a metric, and, hence, a unique affine connection compatible with that metric. Hence we cannot, in general relativity, split the affine connection into a flat, integrable part and a remainder, taking the flat part as the real affine connection of the spacetime and the remainder as representing gravitational force, as we could in local Newtonian spacetime.[22]

But the argument is fallacious. All that is shown is that if we do split the affine connection into a flat part and a gravitational remainder, then we shall have to adopt the metric appropriate to flat Minkowski spacetime as representing "real" intervals in spacetime, and we must then postulate an additional physical field representing the *"metric" effects of the gravitational field.*

Of course there is a physical difference here from the situation in the Newtonian case. At the very core of general relativity is the equivalence principle which tells us that the gravitational field locally acts like an accelerated coordinate frame — on electromagnetism and on metric measurements as well as dynamically. Combining this with the metric effects so easily predictable on frame transformation in special relativity one rapidly comes to the conclusion that the gravitational field must not only "force" particles and bend light waves but influence rods and clocks, i.e., metric measurements, as well. The well-known gravitational red-shift arguments, for example, show this clearly.

So, of course, if we are to try to give a flat spacetime rendition of general relativity we will have to introduce a gravitational field which not only has dynamic effects, but which has electromagnetic and *metric* effects as well. This is what the argument from the necessary compatibility of affine connection and metric shows us. It certainly does not show us that a splitting of the affine connection in general relativity into flat part and gravitational remainder is impossible.

Indeed, the plausibility of such a splitting in general relativity is dependent on contingent facts about the universe just as it is in the Newtonian case. That is, in the island universe case we have a natural splitting: the one which takes the inertial frames of observers far from matter as the real, flat inertial frames of the flat spacetime. In the cosmological models which are filled with matter we have no such natural flat rendition of general relativity. So we

22. Ibid., 355.

would have to let any free falling, nonrotating observer extend his local inertial frame to be the flat global frame (with very peculiar results) thereby violating our cosmological principle. But in either case we can certainly *formally* give a flat spacetime version of general relativity, by appropriately adopting simultaneously a flat metric tensor and its associated compatible affine connection, so long as we refuse to say anything about just what measurements *really* measure the metric or affinity and hold to the ordinary physical measurements as reading off the original "metric" and "affine connection" which represents the "sum" of spacetime and gravitational field effects.[23]

The other arguments of E&F upon examination once again show us only that the scientific details of the way gravity functions either as a component of a curved spacetime or, alternatively, as a field impressed on a flat spacetime, are different in general relativity and local Newtonian spacetime.[24] Of course they are. Basically what they show is that the choice is not curved spacetime or flat spacetime plus gravitational force, but, rather, curved spacetime or flat spacetime plus gravitational *field*. But none of this is relevant to the claim that the issues of conventionality and the related issues of substantivalism versus relationism have any fundamental philosophical difference in the two cases.

VI

The particular scientific theory we adopt does, indeed, tell us just what the conventional alternatives, if there really are such, must be. But the allegation of conventionality, the approaches to denying its reality or obviating it by ontological retrenchment, and the general issues of substantivalism (or, more generally, of commitment to theoretical entities) are philosophical issues which rest upon deep-seated epistemic grounds. They rest upon the difficulties which inevitably occur when we posit entities which are not open to direct epistemic access, and the equally disturbing conclusions we are led to when we try to ease our pain by eliminating such commitments.

23. For more on this, see Havas, "Four-Dimensional Formulations," 961–962, and the references to the work of Rosen cited therein. See also S. Deser, "Self-Interaction and Gauge Invariance," *General Relativity and Gravitation* 1 (1970): 9–18, and Misner, Thorne, and Wheeler, *Gravitation,* 424–425, for more on general relativity and flat spacetimes.
24. Earman and Friedman, "Meaning and Status," 355–356.

The study of the alternative spacetimes in which we can formulate Newtonian and general relativistic physics can be very illuminating philosophically. For these spacetime theories provide extraordinarily perspicuous test cases and examples for our philosophical arguments. But the hope that we can resolve the philosophical issues by getting clear enough on the science is, I believe, as vain as the hope that we can avoid the hard empirical work of science by thinking hard enough philosophically.

7. Semantic Analogy

Even a three-year-old child can understand a story about "people too
little to see". . . —

> Hilary Putnam, "What Theories Are Not"

If one has to imagine some one else's pain on the model of one's
own, this is none too easy a thing to do: for I have to imagine pain I
do not feel on the model of the pain which *I do feel.—*

> Wittgenstein, *Philosophical Investigations*

*This chapter explores the contrast between two distinct accounts of the way
in which meaning accrues to those terms of a theory which refer to unobservable entities and properties.*

*One account, by far the most popular, credits meaning accrual to theo-
retical terms as being solely the result of the place the term occupies in the
theory as a whole. Thus, observational terms excluded, what a term means
is fixed entirely by means of the place which it occupies in the theoretical
structure. This account, while intuitively plausible in many cases, and
while quite adaptable in leading to such things as systematic accounts of
theoretical equivalence, has a tendency to drive its proponents (if they are
conscientious) to instrumentalistic or other antirealistic accounts of theoretical reference, etc.*

*An alternative account, semantic analogy, is rather harder to construe.
Basically, the claim is that some terms referring to unobservables gain
meaning by the role they play in a quite different context of referring to
observable entities and properties. Crudely, 'red' means the same thing in
"Red particles too small to see," as it does in "Visible red apples." The
idea is that meaning accrues to a term from its use in describing the
"directly observable" and carries over intact to the term in its theoretical
role.*

This essay argues that while such meaning accrual through semantic analogy is difficult to accurately characterize, and that when allowed to work unrestrained it leads to paradoxical and unacceptable results, it is intuitively plausible as an account of the meaning of theoretical terms in certain contexts. Further, some aspects of the realist's account of scientific explanation and of the very point of theories become clearer from the semantic analogy point of view than they are from the point of view of the more orthodox "meaning as place in theory" account.

I

FACED with the difficulty of rationalizing the intuitive nondeductive principles of inference we employ as guides to a transcendent truth one may be tempted, in the vein of Peirce, to offer a reductive account of what truth is, taking it to be that which is, if only in the "long run," that which is warrantedly assertible.

Similarly, faced with the apparently intractable problems one encounters when one takes meaning to be characterized in terms of truth-conditions, there is a strong temptation to offer an account of sense which ties it primarily to the conditions under which we are warranted in asserting and denying a proposition. Such a verificationist account of meaning is supported even more strongly by the doctrine of meaning as use, and by arguments to the effect that learning the meaning of an expression can only be learning when it is appropriate to use it or to accept or criticize another's use of it; and, hence, that what is learned when meaning is grasped must be, primarily, these very conditions of acceptancy and rejection.

That an adequate theory of meaning will invoke the conditions under which we are warranted in accepting and rejecting an assertion at some points in its structure is something many will accept. But a very fundamental question is this: Must an account of the meaning of an assertion be an account of the grounds for accepting and rejecting it for *every* assertion whose meaning is to be characterized? Or might the situation be, rather, that for some assertions one must, to give their meaning, stipulate their verification and falsification conditions, but this need not be done assertion-by-assertion for every proposition accepted as meaningful? Could we not, in some way, *project* an understanding of meaning throughout language, using its structural features as the guide to correct projection, from some fundamental basis to which meaning accrues in a manner acceptable to a verificationist?

The basic idea is simple, if, admittedly, terribly vague. For those propositions whose truth-conditions are accessible to us, meaning is given by stipulating the conditions of warranted acceptability and rejectability. For these propositions giving the verification and falsification conditions *is* giving the truth-conditions and, surely, on anyone's view of meaning, fully characterizing the meaning. But for those propositions whose truth-conditions are, somehow, "inaccessible" to us we are blocked from this route. The solution is not to force us to reject our intuitive picture of the proposition as having inaccessible truth-conditions, offering a verificationist revision of just what the meaning of the proposition "really" is; but, rather, to offer an account of how we can grasp the meaning of these propositions without stipulating for them *directly* conditions of warranted assertability and rejectability. Instead we offer an account of how an understanding of these propositions arises indirectly out of their structural relationship, intralinguistically, to the propositions whose truth conditions are open to us.

It is one version of such a theory that I intend to look at here.

II

It would be useful here to make some brief remarks about the overall structure of various aspects of the traditional verificationist program. While everyone is familiar with many of the objections in detail which apparently vitiated these programs, some remarks about the general structure of the alleged failures will be important here.

The positivists tried to present both criteria of meaningfulness and a theory of meaning. The criteria of meaningfulness were usually framed in terms of *sentences,* principles being offered to determine when a proposition was meaningful in terms of its bearing some appropriate logical relationship to the distinguished class of observational, and hence prima facie meaningful, propositions. Of course there is also a tradition of attempting to apply a criterion of meaningfulness for *terms* as well. Theories of meaning were presented in the form of a theory of how *terms* accrue meaning on the basis of an antecedently given observational vocabulary. That a theory of meaning would be given as a theory of term meaning seems an inevitable consequence of componentialism, i.e., of the realization that a theory of meaning must proceed by terms unless an infinite number of distinct sentences are to have meaning assigned to them one at a time. Curiously, not much attention was paid by the verificationists to the fact that even if we could characterize the class of meaningful terms, and specify their meanings, we would still need a theory which tells us when a given combination of terms into a sentence is meaningful and which tells us how to generate the meaning of the whole sentence out of the meanings of the component terms. This will be essential for us later.

The history of the verificationist programs is a history of progressive weakening. First, only sentences strictly verifiable and strictly falsifiable are considered meaningful. But then vast numbers of prima facie meaningful propositions of science are declared metaphysical or, at least, unscientific. So the criterion is weakened in one direction or another. Well known is the direction in which one allows the use of auxiliary sentences in testing a sentence for meaningfulness by exploring its deductive connections with observational sentences. But then it became hard to formulate a criterion which excludes any sentence at all from the realm of the meaningful. Alternatively, one can weaken the criterion by allowing inductive or probabilistic relations between observation sentence and theoretical sentence rather than insisting on a strict deductive relation between the two. Or one could move in both directions simultaneously. If one chooses to deal with terms rather than sentences, a parallel process of weakening is observed as the authors move from operational or coordinative definitions, to reduction sentences, to the holistic notion of partial interpretation. Here again, by allowing inductive connections between observational consequences and theory to legitimize a theory, and hence to give meaning to the terms appropriately appearing in it, a weakening in the other direction is invoked as well.

It is well known that the weakening of the original strict verificationist criteria cannot be accomplished without severe difficulties infecting the program. Once allow as meaningful an assertion whose truth-conditions cannot be fully stipulated in terms of the observational vocabulary and it becomes hard to see how any sentence or term can be excluded from the realm of the meaningful. This is most obvious when the move toward holism is made, that is when "auxiliary sentences" are allowed in testing a sentence for meaningfulness or when "partial interpretation" of terms replaces strict operational definitions. What we must be clear about, though, is that the failure to produce a definitive verificationist criterion of meaningfulness is not the result of mere technical difficulties, but is, instead, the result of very serious difficulties the approach faces as soon as it abandons the most naive and rigid standards of strict necessary and sufficient truth conditions framed in the observation language.

Each direction of weakening, the *holistic* which allows the invocation of auxiliary sentences when testing a sentence for meaningfulness or the invocation of other theoretical terms when testing a term for meaningfulness, and the *criterial* which allows the use of inductivist or probabilistic connections between the theoretical and the observational to serve as the test for meaningfulness, suffers its own characteristic difficulties. The real force of these difficulties only becomes apparent when one moves from the comparatively easy task of specifying a criterion for meaningfulness to the much more difficult task of offering a full-fledged theory of meaning for theoretical discourse.

The holistic case is the more familiar. Once take whole theories as the appropriate bearers of meaning, and, in the domain of inference, take responsibility to the phenomena as the basic criterion of warrant for their acceptability (allowing, possibly, as well simplicity, methodological conservatism, et al. to also play a role), it then becomes hard to see how either independent non-theory-relative meaning is to be assigned to any sentential atom of the theory, or independent reference and extension to that sentence's component singular terms and predicates. On the inferential plane it is hard to see how independent warrant is to be assigned to sentences one at a time. This difficulty shows itself most clearly in the familiar feature of duality. Given a theory which adequately saves the phenomena we can also cook up innumerable others, superficially inconsistent with the given original theory, by simply compensating for variation in one place with a saving variation elsewhere. On the level of inference such duality leads to skepticism, apriorism, or their not-particularly-intelligible relative, conventionalism. But, much deeper, is what duality leads to on the level of *meaning* theory. Either directly, as in Eddington, Reichenbach, et al., or more obscurely on the part of some recent authors, duality on the theoretical level leads us by a smooth chain of argument to a denial of genuine (or, perhaps, unrelativized) meaning (and reference) to the sentences of the theory taken one at a time or to their components. Truth is denied to the sentences individually, but said to be applicable to them only relative to the remaining body of theory; and the reference of the terms and predicates of the sentences is again now only "relative" to the remaining theoretical context. Commonly (and plausibly) we are led by such arguments to the dilemma of either retreating to the reductionist analysis of the meaning of theoretical assertions which the strict verificationist had in mind in the first place, or, instead, to a denial of reductionism which succeeds in finding a role for the individual theoretical sentences by, in its meaning theory, denying them meaning properly so called at all—that is, by a retreat to instrumentalism for the theoretical apparatus.

The loosening of strict verificationism which proceeds by way of replacing entailment relations with relations of warranted assertability, inductive probability, etc. also faces serious problems. Here the fundamental difficulty is not of holism. In the case where verificationism was weakened by allowing in auxiliary sentences in tests of meaningfulness and by taking meaning as role played in total theory, the difficulty is in reconciling an empiricism which fundamentally takes whole theories as the units of meaning with an attempt to preserve notions of truth, reference, etc. distributed "atomistically" over individual sentences and their component referring expressions. In this new case the problem is trying to reconcile the new theory of meaning appropriate to this weakened criterion of meaningfulness with the basic intuitions we have connecting the notions of *meaning* and *truth.*

It is almost a truism to claim that a stipulation of the meaning of an assertion requires *at least* a stipulation of its truth conditions. This might not be enough, as we know, because of all the familiar arguments to the effect that logically equivalent assertions need not be synonymous, etc. But, surely, one argues, saying under what circumstances an assertion is true and under what false constitutes a *necessary* component in specifying its meaning. Then one goes on to account for the meanings of the components of the sentence in terms of their contribution to the sense, i.e., the truth-condition stipulation, of the whole.

But if we are to be allowed to introduce novel assertions into our language on the basis solely of their inferential relations to already understood assertions, where these inferential relations are less than deductive and so are not sufficient even to characterize the truth-condition of the assertion so introduced, then what are we to make of this intuitively obvious claim that the meaning of an assertion is a stipulation of the conditions under which it is true and under which it is false?

The theory which takes the fundamental primitive in meaning theory to be warranted assertability rather than truth faces a dilemma. Either it reserves a place for a limited class of assertions whose meaning is given by our recognition of the truth-conditions for them (observation sentences, perhaps) or it does not. In the former case those assertions not in the privileged class seem to have such an attenuated sense of meaning left over for them (in comparison with that rich, full sense of meaing possessed by the members of the privileged class whose meaning is their recognizable truth-condition) that we are easily led to an instrumentalistic account of their role in our conceptual scheme. Their meaning doesn't refer to their truth-conditions, their singular terms don't have reference in the full sense, nor their predicates extensions. Why then take their apparent ontological commitments, etc., seriously at all?

The alternative is to hold that the meaning of *all* assertions is given by their conditions of warranted assertability and deniability and that the recognizability of truth-conditions never plays a role in the comprehension of meanings. Such a flat-out anti-truth-condition theory of meaning is hard to criticize, expecially since, despite Dummett's suggestive work, no such theory has even really been made available to us for our inspection. Dummett suggests to us that the adoption of all-out verificationism will, at the very least, require a revision in our logic, from traditional to intuitionistic logic or some variant thereof. Far more serious is the full antirealism which one will then seem to be committed to. If all sentences have as their meaning something specified in terms of conditions of warranted assertability and deniability; if for all of them meaning is disconnected to the notion of truth-condition, and, hence, for their components, meaning is disconnected from the orthodox notions of reference and extension, what will be left of the connection between language and the world at all? It is one thing for Wittgenstein, advocating a verificationist account of mentalistic language, to tell us that a nothing would do as well as a something so far as the meaning of 'pain' is concerned. It is something else again to adopt such a cavalier attitude toward being in general.

The parallel here with the case of inference is clear. Once adopt a foundation language of propositions warranted by direct access to the states of the world they describe, and skepticism about the truth of the remaining sentences, believed on such far more slender warrant, rears its head. Deny the existence of foundation statements altogether and one quickly finds oneself on the slide from a coherence theory of warrant toward a coherence theory of truth. What we see here is that on the level of the theory of meaning, and even without the coherentist's holism, a doctrine which eliminates truth-conditions altogether in favor of conditions of warranted assent and dissent soon loses any ability to account for the connection between language and an independent world of things, properties, and facts. Perhaps, once again, the world is well lost, as Rorty would have us believe, but it is a radical step to take and one we would do well to hesitate before taking. Incidentally, this dilemma of either reserving a special class of observation sentences or else treating all sentences on a par gives rise to similar difficulties for the holist as well.

III

Given the difficulties of sustaining a realistic interpretation of terms and sentences referring to unobservables on either the holistic or verificationistic accounts, it would seem prudent to at least explore any prima facie viable alternative which presents itself. One such alternative has particular appeal, since the model of understanding it presents is one which seems to us pretheoretically (rightly or wrongly) to have plausibility to it. The view is this: Basic sentences describing epistemically accessible states of affairs are learned by us by the presentation to us of the appropriate states of affairs. The meaning of the assertion is then given by association of it with its directly available truth-conditions. But such basic sentences are made up out of components and, indeed, on the model of any adequate theory of language based on a componentialist semantics (and what other kind could possibly explain our mastery of a potentially infinite collection of novel sentences?) the meaning of the whole sentence must be understood as being generated out of the meanings of the components and their grammatical assemblage into the whole sentence. But then we can understand, once again on a componentialist theory of meaning for the wholes as generated out of the meaning of parts, novel sentences constructed out of these very same elements, the elements first introduced to us through their role in sentences whose truth-conditions are open to our direct inspection. But these novel sentences may describe states of affairs not open to our observational capacities. Hence, on a basically verificationist theory of meaning, one can understand how it is that one can have an understanding of sentences which describe states of affairs whose truth-conditions cannot be "presented" to us. But, on this account, the meaning of sentences is supposed to be given individually, rescuing us from the dualities of holism and its conventionalist consequences, and the meaning of the novel assertion is connected to its truth-condition, and not to conditions of warranted assertability and deniability, and, hence, the new approach is supposed to give us genuine reference for the components of the novel assertion and genuine truth for the assertion as a whole.

That such an account is fraught with difficulty is, I think, obvious. But that it is one many people intuitively accept on a pretheoretical level is also clear. Knowing what 'red' means from such cases as 'This is a red apple' I have no difficulty in understanding what it is for something to be a red object too small to see. Knowing what pain is from my own case I understand (*pace* Wittgenstein) what it is for you to be in pain. Remembering the famous (notorious?) argument from analogy to the legitimacy of belief in your mental states on the analogy with my own, it becomes not implausible to call such a theory as the present one a theory of *meaning* by analogy. The general structure of the theory here looks like this: Initially meaning accrues to whole sentences by an association of the sentence with its (conventionally determined) truth-condition, which "state of affairs" is directly open to epistemic access. Meaning then accrues to sentences in the language whose truth-conditions are not so accessible by a process of decomposition and reassembly which proceeds by an understanding of the way in which the meaning of whole sentences is formed out of the meaning of their parts.

I don't think that even that loose characterization will do for all those cases which one might call projection of meaning by analogy. For example, it is sometimes claimed that quantification over an infinite domain is understood "analogically" from quantification over finite domains. Here the "projection" of meaning by analogy does not seem obviously assimilable to the notion of sentential decomposition and reassembly.

As a first step at exploring the possibility of such a notion of "semantic analogy" playing a respectable role in an overall semantic account of meaning, let us first look at a number of objections to such a theory of meaning accrual. These objections are basically of two kinds: from cases and from general principles.

The refutation of semantic analogy by cases goes like this: The existence of obviously nonsensical sentences made up "grammatically" out of parts which are plainly meaningful when they appear in other sentences shows us that projection of meaning by decomposition and reassembly is illegitimate. There are two kinds of cases, those of "ordinary language": 'It is five o'clock on the sun'; and those of science: 'Two spatially separated events are simultaneous', 'This electron is green'.

Now of course we could deny the meaninglessness of the sentences adduced. But even if we accept them as meaningless, their existence doesn't show that projection of meaning by decomposition and reassembly is *always* illegitimate. It does show that it *sometimes* is. The existence of such cases (very familiar from Wittgenstein who wishes to break the alleged stranglehold of the "false picture" of meaning by analogy has on us) does indeed show us that the advocate of the legitimacy of meaning by analogy has his work cut out for him. For he must make at least the first steps toward a *theory* of semantic projection which will delineate the range of legitimate projection of meaning by analogy. Such a theory might at least initially proceed by a theory of "semantic categorization" of terms, marking out by means of construction rules legitimate from illegitimate combinations in terms of the semantic categories into which terms are placed. Such a theory would not, I believe, work in general. In any case, it could be but the most rudimentary intermediate step in our theory, for the very notion of semantic categorization itself calls out for explanation.

Here, as elsewhere, the parallel of the theory of meaning with the theory of believability is interesting. As we now know, the believer in inductive inference has his work cut out for him as well. Even prior to the establishment of a final "justification" for inductive reasoning, he must first give us at least a rudimentary description of what inductive reasoning consists in. Even on the very naive model of inductive generalization as induction by enumeration, the existence of unnatural classifications of things (with 'grue' as the extreme case of an illegitimate classificatory term) shows that there is a fundamental problem first to be faced in trying to delineate the legitimate from the illegitimate cases of inference and to somehow rationalize (or at least explain) the boundary between the two.

Over and above the "refutation by cases" (the form of refutation which appears very clearly in the relevant portion of Wittgenstein's

Investigations) one can find attempts at more "theoretical" refutations of the view that meaning can accrue by the analogy of epistemically unaccessible to epistemically accessible states of affairs.

One such refutation is that of Dummett.[1] Basically, the line of argument is this: The theorist who believes that meaning can accrue by analogy is basically a "realist" in semantics. That is, he takes the meaning of an assertion to be its truth-conditions. But then how do we understand the meaning of an assertion? Only by "grasping" its truth-conditions. What can this amount to? Either being able to stipulate those truth-conditions in some nontrivial verbal way (say by offering a reductive analysis of the meaning of the assertion) or by knowing what it takes to *determine* the truth or falsehood of the assertion. Now consider those assertions whose truth-conditions outrun the limits of epistemic accessibility. The "meaning by analogy" theorist envisages someone who can determine the truth or falsity of such an assertion by means "analogous" but not identical to those we employ. It is such an extended notion of verification, he claims, which is sufficient to give meaning to these assertions, and reference to such an extended method of verification which, indeed, specifies what the meaning of these, in actual fact, unverifiable and unfalsifiable, assertions is.

Now, Dummett continues, where the analogy of the new extended method of verification to our usual methods is weak or nonexistent, the analogist admits that meaning cannot accrue in this way. Hence, while it might be legitimate to think of meaning accruing to quantifications over infinite totalities by envisaging someone who can "survey" an infinite class as we survey a finite class by exhaustive enumeration of cases, none but the most ardent analogist is likely to be satisfied with a theory of the meaning of modal assertions which gives us a verifier with the extended capability of surveying the contents of other possible worlds.

But, Dummett says, these distinctions are only of psychological importance, explanatory only of why the analogist *thinks* that he has a theory of meaning even in his best cases where the analogy is clearest. For even in those cases he is viewing the meaning of an assertion not as we use it but as his superpowered verifier uses it and, hence as far as a theory of meaning of assertions in our lan-

1. M. Dummett, "What Is a Theory of Meaning? (II)," in G. Evans and J. McDowell, eds., *Truth and Meaning* (Oxford: Oxford University Press, 1976), 67–137. The argument discussed above is on 98–101.

guage does, he is *mis*representing the use of the assertion in language and *mis*construing its meaning.

I think, though, that this refutation of what Dummett calls (curiously) "strong realism" isn't conclusive against the analogist's case. Even if we take the analogy to be, as Dummett suggests, one between our actual methods of verification and those of an imagined super-being, is it so clear that the analogist is blocked in claiming that the presentation of such an analogy is enough to give us a grasp of the meaning of the assertion in question? The claim that this must be a misconstrual of the meaning as the method of verification described is not that actually employed by us is persuasive only if one has already accepted the verificationist's identification of the meaning of an assertion with its method of verification as it is actually used, which is, of course, exactly what the analogist is anxious to deny.

Furthermore, is it so clear that it is an analogy of methods of verification that the analogist really has in mind? To be sure, he identifies the meaning of an assertion with its truth-conditions and grasping the meaning with "grasping" the truth-conditions. But why should he agree that this can only amount to grasping that method by which we, or at least some analogically more powerful extension of our selves, would determine the truth or falsehood of the assertion? Once again that is only plausible if one is a verificationist to begin with. Most likely the analogist's position is, rather, something like this: For the basic assertions, those whose truth-conditions are directly accessible to us epistemically, we learn their meaning by associating them directly with the presented truth-condition. Of course, since directly being aware of the existence of the truth-condition is the primary and most conclusive warrant we could possibly have for believing in these propositions, one could also identify the meaning of the assertion with the primary grounds for warranting belief in it — without committing oneself, of course, to a verificationist theory of meaning. For those assertions whose truth-conditions are not open to our epistemic access, we understand their meaning by grasping their truth-condition. We do this latter by seeing the analogy of the *truth-condition* of this new assertion with the truth-condition of some one or more of the basis assertions. But it is an analogy of *truth-conditions* which is relevant, not of methods of verification at all.

An example might make this clearer. How do I know the meaning of 'He is in pain'? Well, from my own case I know the meaning of 'I am in pain'. By decomposition (structural understanding of the latter assertion and how its components function together to make up a meaningful whole) I know what 'is in pain' means and by reassembly I know what 'He is in pain' means. Isn't that the analogist's argument? It is not that I imagine someone who, unlike me, can immediately experience everyone's pain, nor even that I grasp the meaning of 'He is in pain' by seeing the analogy between his directly experiencing his pains and my experiencing mine, and then relying upon the association of meaning with method of verification to think I now grasp the meaning of 'He is in pain'. No. It is that the meaning of 'He is in pain' is given by its truth-conditions, and *these* I understand by their analogy with the (directly epistemically accessible) truth-conditions of 'I am in pain'. It is something like a *state-of-affairs* which I am supposed to "grasp" on the analogy with a directly experiencible state-of-affairs that is at the heart of the analogist's claim seen from this point of view. Of course this is not a theory of meaning and it isn't clear that there is any coherent theory which can "back up" this initial "model" or "picture." We will return to this problem shortly.

I V

I have no intention here, of course, of even making reasonably coherent suggestions about just what a theory of meaning would be which would allow for the embedding in it of a theory of legitimate projection of meaning by means of analogy. Rather, I will just note four aspects of the problem of meaning by analogy which are of some importance, aspects which will come under discussion, I think, no matter what the general meaning theory is to be which allows for analogical projection of the comprehension of the meaning of assertions.

Skepticism

If we agree that we can grasp the meaning of propositions on the basis of their analogy with previously understood propositions, we must realize that we have opened ourselves up to the possibility of meaningful assertions relative to which no grounds of warranted assertability and deniability are forthcoming (or, perhaps, even could be forthcoming). This is not to say that the believer in semantic analogy *must* be committed to the comprehendability of at least some "undecidable" propositions, but only that he might be.

In this way he surely differs from the verificationist. If the very meaning of the proposition is fixed by the warranting conditions for it, one need never fear that one will come upon a meaningful assertion for which warranting conditions are not forthcoming. Probably, also, in this the believer in semantic analogy differs from the holist as well, although it isn't as clear in this case as it is in the case of the verificationist account of meaning. For the holist the meaning of the assertions not asserting the existence of conditions directly open to epistemic access is fixed by the role the assertion plays in a total theory. But the total theory must be responsible, presumably only to the empirical facts which could serve to confirm or disconfirm it. Now if we take it as a sufficient condition of theoretical synonymy (two theories being equivalent in the sense of "saying the same thing") that the two theories have a common body of observational consequences, then once again skepticism has been undercut. We need not fear the existence of alternative theories which cannot be judged relative to one another because they have their empirical consequences in common, for in the case of common consequences we just assimilate the theories together by characterizing them as synonymous. And of course the holist will not countenance our asking the question of how we test individually the assertions of the theory for acceptability, since it is just this consideration of the assertions "one at a time" which he will not allow.

But if it is true that both verificationism and holism amount to a denial of realism for the assertions whose truth-conditions lie outside the range of direct epistemic accessibility, then it is hardly a surprise that the semantic analogist has greater concern than they over the problems of skepticism. for he takes these "theoretical" assertions to be true and false, as are the "observational," and their components to have genuine reference, extension, etc. And yet he freely admits that the truth-conditions for the assertion are not the sort we can simply observe to either be or not be the case. If I take it that 'He is in pain' is an assertion of the existence of a mental state of another, not open to my epistemic access, then of course I will worry about how I can ever *really* know whether or not someone is in pain in a manner which will not bother at all someone to whom a nothing would do as well as a something as far as the meaning of 'pain' was concerned.

The Limits of Meaning by Analogy

To just what extent can we project meaning out of our basic sentences and on to new assertions by means of semantic analogy? Perhaps, if we take basic sentences as reporting the contents of immediate subjective awareness, we could project meaning onto assertions about the external world. At least Locke thought that we could, although Berkeley's familiar criticisms make us pause even here. Perhaps, using a projection of meaning by analogy, we can understand sentences attributing mental states to others on the basis of our understanding of mentalistic assertions about our own private mental states. Although, again, Wittgenstein would have us doubt even this. But to what extent can we support, for example, a realistic interpretation of theoretical assertions in science by a program which initially attributes intelligible meaning to these assertions on the basis of their semantic analogy with assertions of the observation language assumed to be understood?

Perhaps it isn't too unreasonable to view bacteria as, among other things, tiny objects too small to see. Even for molecules the analogy is "good enough" to allow us to at least plausibly maintain that we understand the kinetic theory of gases on the analogy with a box filled with rapidly moving billiard balls. But when we get to electrons, quarks, photons, or worse yet, virtual intermediate massless bosons, charm, etc. what earthly use can we make of analogy as a source of meaning?

I think there are two things that can be said. First, even in some pretty recherché cases there is still a certain amount of theoretical predication going on which can at least plausibly be argued to be understood in the analogical way. Even if virtual particles are sufficiently remote from dust particles in their properties that to speak of them as particles at all is just to mislead, still they do have, perhaps, spatiotemporal location, momentum, energy, etc., and perhaps these can be understood on analogy with those features when predicated of objects of experience. Second, insofar as the features of things cannot be understood on any analogy whatever with features of the elements of experience, we can, at this point, always retreat to an instrumentalistic denial of full meaning to the predicates altogether.

Electric charge is just the disposition to behave in certain ways under certain test conditions; the psi-function of quantum mechanics is just an "intervening variable" connecting test conditions and results of measurements, both characterized in the old observational vocabulary, but having no independent meaning or reference of its own, etc. Witness Bohr's insistence that our body of *real* concepts will always be restricted to those familiar to us from everyday experience, the rest of the theoretical apparatus to be understood only functionally relative to assertions framed in this original vocabulary.

Now it may very well be that the disposition on our part to hold to a realistic interpretation of theoretical assertions when they are framed in the familiar vocabulary of observational experience, and retreat to instrumentalism only when terms are invoked which can't be understood analogically on the basis of observational predicates is a mere psychological prejudice on our part. Certainly this would be the claim of the orthodox verificationist and the holist. But I think it would be as premature to dogmatically accept such a claim and reject our "intuitions," vague as they may be, as it would be to rashly accept the Humean claim that our belief in the uniformity of nature is but a reflection of psychologically explicable prejudice and not a rational and rationalizable approach to an attempt to know what goes on in the world.

Legitimate vs. Illegitimate Analogy

As was noted earlier, a primary assault on the very notion of semantic analogy proceeds by attempting to show that the principles the analogist uses to allow the projection of meaning from sentence to sentence — basically, I have claimed, the principle of decomposition and reassembly — are such that they lead to the claimed intelligibility of prima facie meaningless utterances ('It is five o'clock on the sun', 'The number three is thinking of Vienna', etc.). Now obviously the analogist can reply that he doesn't accept *any* decomposition and reassembly of sentential components as guaranteeing a transfer of meaningfulness from the original sentences decomposed to the newly reassembled utterances. There must be some restrictions on the range of decomposition and reassembly.

But how these restrictions are to be formulated is going to be a matter of some difficulty for the analogist. I do not believe that any naive attempt in terms, say, of a categorization of vocabulary into semantic classes and an attempt to delimit legitimate rules of assembly in terms of these classes will do the job. Even the most obvious rules we think of will meet a verificationist challenge, we might note. Mere substitution of indexicals is itself dubious, according to the verificationist, for, according to him, I *don't* understand 'He is in pain' on the basis of 'I am in pain' even though both indexicals refer to persons. Again, if some interpreters of relativity are right, Einstein showed that we were mistaken in thinking that we could understand the simultaneity relation when predicated of events at a distance from one another, just because we understood it when predicated of spatially coincident events. Now the verificationist might be wrong about these cases, but how is the analogist to convince him without a full-fledged theory of the legitimacy of analogical projection of meaning and without an analysis of the limits of legitimate meaning projection by analogy?

The comparison with the verificationist program is interesting here. The verificationists provided both a criterion of meaningfulness, and, much more interesting, a doctrine about meaning. The two fit together in an obvious way: to specify meaning is to give the conditions of warranted assertability and deniability. In the absence of such conditions, the assertion is meaningless. What is the analogist's theory of meaning to be, and how, relative to it, are the limits of legitimate projection of meaning by analogy to be characterized?

A Theory of Meaning

We now come to the most fundamental problem with the doctrine of the projectability of meaning by analogy. Into just what theory of meaning and the knowability of meaning is it to be inserted as a viable part?

The theory is one which takes as fundamental in meaning theory truth-conditions and our "grasp" of them. To understand the meaning of a proposition is to know what it asserts, i.e., to understand what it is for it to be true and what it is for it to be false. We can "grasp" a truth-condition by having it directly accessible to us (we are immediately aware of the conditions asserted to be the case for the "observation sentences") or we can "grasp" a truth-condition by seeing its analogy to a truth-condition already comprehended by us, that is by "constructing" the new truth-condition out of components available to us from their role in antecedently comprehended truth-conditions according to structural principles of constituting truth-conditions out of objects and properties also familiar to us from understood cases.

But what all this means is far, far from clear. It smacks all too much of a "picture in the mind" theory of meaning, where comprehension of assertions is viewed in the manner of an association of sentences with pictures and of novel sentences with new pictures made up out of pieces cut from old ones and glued together in novel ways. This won't do, of course, as a theory of meaning. But just what does the analogist have to put in its place? In particular, what does he have which is informative over and above the trivialities which tell us that to understand the meaning of a sentence is to grasp its truth-condition, which is, after all, just to know what the sentence means? One thing "grasping a truth-condition" cannot mean, of course, is knowing how to determine the truth or falsity of the proposition. For, according to this view and in direct conflict with verificationism in any of its forms, we can perfectly well know the meaning of a proposition and yet have no idea whatever how one could possibly determine whether it be true or false. And it can't mean simply knowing the necessary and sufficient conditions for the truth of the proposition.

Despite the difficulties such a theory of meaning and the comprehension of meaning obviously would meet, it still might be worth some effort to find such a theory. To convince you of that let me end with a (familiar) little parable. Exploring a distant planet, we come upon some entities. Utterly unlike any living creatures we are familiar with, we still find their behavior complex enough to offer an account of it in terms of posited internal functional states. Let us call one such state being in 'grinch'. Our complex functionalist theory sometimes leads us to say "That one is in the grinch state." After a while a particularly imaginative alien-anthropologist suddenly sees close formal parallels between the functional organization of the internal states of the creatures and the "program" of functional organization of human mental states. In this formal parallel, 'being in the grinch state' is located just about where 'being in pain' is located in our description of the mental life of men. So, speculating, he says: "Perhaps being in grinch is being in pain."

Now we are not concerned with whether or not he is right. Or even if he ever has any good reason for believing he is right in identifying grinchness with pain. The point is, do we not believe that in coming to entertain the proposition 'That thing is in pain' he has come to entertain a novel proposition different from merely believing 'That creature is in the grinch state', and, indeed, a proposition he can understand only because he knows what pain is (presumably, but not necessarily, from his own case). But his means for establishing whether or not a creature is in pain are, of course, the very conditions he uses for establishing whether or not a creature grinches. Yet the proposition he now takes warranted by "grinching behavior" is not merely that the creature is in a grinch state, but that it is in *pain.* If you accept that a novel proposition has been entertained by him, then you accept the claim that there is more to meaning of an assertion than its conditions of warranted assertability and deniability. And if you accept the claim that this additional element is a grasp of meaning which comes about from understanding the meanings of the components in the novel assertion and from understanding the role they had in previously understood assertions, then you ought to pursue an analogical theory of meaning projection, elusive as such a theory might be.

8. Incongruous Counterparts, Intrinsic Features, and the Substantiviality of Space

Kant once argued against the relationist account of space, basing his argument on the claim that relationism could not account for the existence of incongruous counterparts, such as pairs of right- and left-handed gloves where the gloves were, aside from their handedness, identical.

Recently Nerlich has argued against relationism, not from the existence of incongruous counterparts, but, rather, from the fact that whether or not two objects are incongruous is a feature not only of the objects themselves and their nature, but of the global properties of the space in which they are located. He argues that the dependence of such facts as congruity or incongruity on overall properties of space itself is persuasive in favor of the substantivalist claim that space itself exists as an entity over and above the material things in it and their spatial relations to one another.

I argue here that nothing in the facts about congruence or its absence, nor in the dependence of congruity and incongruity on global facts about space, refutes the relationist approach to the metaphysics of space. Some details of the facts about congruity and incongruity, and about other features of things which depend upon facts of global topology of the space in which they find themselves, are discussed.

I claim that the relationist, insofar as he can account for spatial features of the world at all, has nothing special to fear from features like that of congruity and incongruity which require reference to the global aspects of the space to have their nature explicated. Further, I argue that any puzzles faced by the relationist will reoccur from the substantivalist point of view.

K ANT argued, as part of his argument that space is an a priori intuition, from the existence of incongruous counterparts (such as right- and left-handed gloves otherwise alike) to the existence of space as an entity over and above the material objects in it and their spatial relations to one another. Peter Remnant and John Earman have argued that Kant's argument is incoherent.[1] Graham Nerlich has recently invoked the dependence of facts about handedness on global features of space to attempt to revindicate Kant's argument.[2] I will argue here that, even taking account of the dependence of facts about handedness on global features of space, noticed by Earman and utilized by Nerlich, there is no good argument against relationism founded on facts about handedness. Or, more precisely, there is no good argument against relationism based on handedness which goes beyond the best standard arguments against relationism which invoke no facts about handedness at all.

I

Remnant and Earman believe Kant's argument incoherent for the following reason: Kant says that, since the internal relationships of parts of left- and right-handed objects are the same, nothing about the structure of these objects could differentiate them. Therefore, what makes them different must be a difference in the relationship they bear to space itself. But if we invoke "space itself," the only account we could plausibly offer to explain the difference in handedness of the objects would be their differential congruity with parts of space itself. But a right- and a left-handed glove would each coincide with some "piece" of space itself; so what makes one left- and the other right-handed? The only answer could be the handedness of the piece of space with which they coincide. But if there is a feature of a piece of space, its handedness, which determines whether an object coincident with that piece of space is right- or left-handed, then why can't the same feature hold of the objects

1. P. Remnant, "Incongruent Counterparts and Absolute Space," *Mind, New Series* 72 (1973): 393–399. J. Earman, "Kant, Incongruous Counterparts, and the Nature of Space and Space-Time," *Ratio* 13 (1971): 1–18. Parenthetical page references to Earman are all to this paper.

2. G. Nerlich, "Hands, Knees and Absolute Space," *Journal of Philosophy* 70 (1973): 337–351. Parenthetical page references to Nerlich are all to this paper. A similar claim that the dependence of orientation properties on global features of space provides a refutation of relationism is made, without argument, in T. Humphrey, "The Historical and Conceptual Relations between Kant's Metaphysics of Space and Philosophy of Geometry," *Journal of the History of Philosophy* 11 (1973): 483–512; see p. 488 n.11.

themselves and serve to differentiate them? And if there is no such "internal" feature of the objects, then how does the invocation of space itself help us?

Earman goes on (7) to maintain that there is such an internal feature of objects, the orientation of their parts. Right- and left-handed objects differ in their internal structure in that their parts have a different orientation with respect to each other. A failure on Kant's part to realize that orientation of parts is just as much an "intrinsic" feature of the objects as, say, size of parts and magnitude of angles between them is the fundamental mistake which led him into thinking that any argument from handedness to belief in space itself as an entity over and above the things in it was either needed or plausible. There is no more puzzle about handedness than there is about any other internal feature of objects which differentiates them. And if there is such a puzzle the invocation of space as an autonomous entity in no way resolves it.

But Earman retrenches a bit (8/9). Suppose our space is globally nonorientable. Then there will be a continuous rigid motion (crm) that brings a handed object into congruence with its incongruous counterpart, even though the two objects cannot be brought into congruence by any local crm.[3] But how could even a global crm change an intrinsic internal feature of an object? So the "intrinsic-ness" of handedness seems questionable.

It is just this theme that Nerlich takes up in his attempt to argue from facts about handedness to the substantiviality of space. At least I take it that Nerlich is arguing for substantivalism. He calls his opponent a "relativist," but never makes it fully clear what constitutes a relativist. His actual claims are:

> Which of these . . . determinate characters [being enantiomorphic, i.e., a member of a possible pair of incongruous counterparts or, instead, homomorphic, i.e., being bringable into congruence with any object of the same size and with the same magnitude of angles between parts] the hand bears depends, still, on the nature of the space it inhabits, not on other objects. The nature of this space, whether it is orientable, how many dimensions it has, is absolute and primitive (345).
>
> . . . space [is] a definite topological entity [and] can only be a primitive absolute entity; . . . *its* nature bestows a character of homomorphism, leftness or whatever it might be, on suitable objects. My

3. Throughout this chapter I assume, without loss of philosophical generality, that the space is one of constant curvature.

conviction of the profundity of Kant's argument rests on my being quite unable to see what the relativist can urge against this, except further relativist dogma (350).

. . . what differentiates a thing which is an enantiomorph from one of its incongruous counterparts is a matter of how it is *entered into the space*. . . .

The idea of entry is only a metaphor, clearly. . . . It is not easy to find a way of speaking about this which is not metaphorical. But a very penetrating but not so painfully explicit way of putting the matter is Kant's own, though I believe it still to be a metaphor. The difference between right and left lies in different actions of the creative cause (351).

I am not sure just what it means to say that space is "absolute and primitive." I believe that the notion of handedness being the result of a "creative cause" is irrelevant to the debate between substantivalists and relationists. I take it that a relativist is one who espouses a relationist theory of space. I think that Nerlich may be right in saying that handedness is a matter of how an object is "entered into space," but I believe that the existence or nature of handed objects is irrelevant to deciding between a relationist and a substantivalist theory of space.

II

Consider two objects. Are they incongruous counterparts, or can they be made congruent by a crm? One might think that there was some "intrinsic feature" of the objects that decided this question. In one sense this turns out to be true, but we must proceed with some caution.

Suppose that the objects are two-dimensional and are both on the same plane. It may be possible to bring them into congruence by a crm that takes at least one of them out of the plane, but not possible by any crm that keeps both confined to the plane. Similarly for objects in our apparently three-dimensional space. If our space really is three-dimensional and if it is orientable, then the objects may be such that no crm can bring them into congruence. Yet *had,* contrary to fact, space been four-dimensional, or if it *had,* contrary to fact, been nonorientable, then a crm *would* have existed which *would* have made them congruent, if the objects are typical "incongruous counterparts."

So, bringability into congruence or its impossibility depends not only upon "the structure of the objects themselves," but upon the structure of space "as a whole," its dimensionality and its global orientability.

Now if space is such that the objects are incongruous counterparts, then we can talk about some "intrinsic feature" of them, that is, some feature of them preserved under all crm's of them, which differentiates them — their handedness. But we must realize that (1) the very existence of such a feature depends upon the overall structure of the space, and that (2) which object has which "handedness" depends upon "how it is situated" in space as a whole.

I think this is true, as far as it goes; but the consequences of this truth must be examined with caution. Notice first that a similar argument can be constructed about a feature of an object that has nothing to do with its handedness.

Consider the objects in diagram (a). Is there a motion that transforms them into the objects in diagram (b) which (1) is continuous and which (2) never brings a bar into coincidence with the circle? Well, that depends. What is the space like in which the objects are situated? If it is two-dimensional, the answer to our question is "No." If it is three-dimensional, the answer is "Yes."

Call the arrangement in (a) "opposite-sided." Call that of (b) "same-sided." Is there an "intrinsic feature" of the object arrangements which is opposite- (same-) sidedness, that is, a feature of the arrangement preserved under all continuous transformations that never bring a bar into coincidence with the circle? That depends upon the space in which the objects are placed. If there is such a feature, what determines which object arrangement has it? The answer is: "How the objects are placed in the space." Perhaps there is some crucial disanalogy here with the left-right case, but I fail to see it. There is nothing mysterious about incongruous counterparts. Many features of a given set of objects, with a specified set of internal relations of its parts to one another, depend both for their existence, and, if they exist, for their nature, on the nature of the space in which they exist and on how the parts of the object are situated in the space.

(a)   (b)

If that is what Nerlich (and Kant?) mean, then they are right. But Nerlich (and Kant) also think that such facts refute the relationist theory of space. They do no such thing.

What is the relationist view of space? It is, at least in the version familiar from Leibniz to Reichenbach, that space is nothing but the collection of actual and possible spatial relations among actual and possible material objects. There are, or may be, some material objects. And there are, or may be, some spatial relations among them. And, spatially speaking, that is all there is. There is no such thing as "space itself considered as an entity" which "exists over and above" the material objects and their spatial relations.

The invocation of *possible* objects and *possible* spatial relations among them is crucial here, just as "permanent possibilities of sensation" are crucial for the phenomenalist. If we wish to be phenomenalists and yet talk about unobserved material objects, then we must, if we are to translate all material-object talk into sense-datum talk without loss of content, tolerate subjunctive as well as indicative sense-datum assertions. Just so, if we wish to talk about places in the world at which no material objects exist, and even more if we wish to be able to talk about spaces totally devoid of contained matter, then, if we are going to translate all talk about "space itself" into talk about the spatial relations among material things, we had better allow talk about possible objects and their possible spatial relations as well as talk about actual objects and their actual spatial relations.

Now the reader might not like the invocation of possibilia, their possible relations to one another, subjunctive or counterfactual assertions, etc. If he finds these totally abhorrent, then he will probably reject the relationist theory of space. Since versions of relationism that eschew such notions are pretty implausible, he may opt for substantivalism immediately.

Alternatively, he may argue like this: Talk about the possible relations among possible material objects is all right, so long as one understands that it is "grounded" in belief in the *actual* nature of *actually* existing substantival space. Just so, we can understand the language of "possible sense-data of possible observers in possible perceptual situations" only because of our belief in actual material objects.

Each of these positions rests on a deep philosophical objection to relationism. The arguments may even constitute devastating objections to the relationist account. All that I wish to claim here is that the following assertions are correct:

(1) Given the full relationist resources, including possible objects and possible spatial relations among them, we can account for all the interesting features of left and right, etc.

(2) Or, more correctly, we can account for these features just as well on the relationist account as we could on any substantivalist account. If there are any "mysteries" about left and right unsolvable on the relationist account, the invocation of space as a substantival entity will be of no use in solving these puzzles.

(3) If there are uneliminable difficulties with the relationist account of space, they have nothing to do with features of the left-right distinction, the enantiomorph-homomorph distinction, dimensionality, or orientability.

(4) The notion of the "creative cause" of the spatial features of objects is of no relevance to the dispute between the relationist and the substantivalist accounts of space.

The notion of space as "absolute" is ambiguous, but, I will argue, notions of orientation are irrelevant to "absoluteness" in any of its senses. It is not clear what sense is to be given to the notion of space as "primitive." Whatever sense we can give to it is such that there seems to be no good argument from the facts about oriented objects to space being primitive in any way that would disturb a relationist.

(1) Suppose that there is a three-dimensional hand. What, from the point of view of the relationist, makes it an enantiomorph, or, alternatively, a homomorph? If it is an enantiomorph, what makes it a left hand or a right hand? Is leftness (rightness) an intrinsic feature of the hand in this case? Suppose that the hand is not an enantiomorph, but is, instead, a homomorph. How can this be the case? If it is the case, is there still some sense in which the hand is still left (or right)? And does leftness (rightness) in this sense now constitute an intrinsic feature of the hand?

According to the relationist, the hand is an enantiomorph if and only if there is a *possible* incongruous counterpart for it. That is, if and only if there is a possible object such that (a) its parts have the same lengths as the parts of the original hand and the same absolute values of the magnitudes of the angles between them, but (b) there is no crm that will bring the hands into congruence. All this is perfectly acceptable relationist talk. No reference is made to "the entity, space, itself," but only to possible objects and possible spatial relations among them.

Now if the hand is an enantiomorph it will be either right- or left-handed. That is, in a world in which there are enantiomorphs the members of a pair of incongruous counterparts are possessed of an intrinsic feature, that is, a feature preserved under all crm's, which is their handedness.

Now Nerlich claims that the handedness of an object is dependent upon "the way it is entered into the space." Is this correct? That depends upon what you mean. If this means that the handedness of an object is dependent upon the spatial relation of its parts to one another, then the claim is certainly true. But then, the triangularity of a triangle depends upon the spatial interrelation of its parts, and so triangularity would also depend upon "how the object is entered into the space."

Is there any interesting way in which handedness differs from triangularity? Well, yes. Handedness, in the full sense, exists only if there is an orientation property of the object which is preserved under all crm's. That is what we mean by the object's handedness in the full sense. But, as we know, the existence of such a feature depends upon topological features of the space as a whole — its dimensionality and its orientability. This is not surprising, for the dimensionality of the space and its orientability determine the class of all crm's and, hence, what is preserved under them. There is no analogous dependence of the very existence of triangularity on the overall topology of the space, and in this sense handedness differs from triangularity. If that is what it means for the handedness of an object to depend "on the way the object is entered into the space," then handedness is so dependent. But that does not mean that handedness is not an intrinsic property of the object in a space in which handedness exists, and there is no good argument against relationism in these interesting topological facts.

But is handedness *really* an intrinsic feature of the object? If you mean by 'intrinsic' a feature of the object which is preserved by all transformations of a specified kind, then handedness may well be an intrinsic feature of an object. For example, if space is an orientable three-space and the object a three-dimensional hand, then if by 'intrinsic' you mean "preserved under all continuous rigid motions," then handedness will be an intrinsic feature of the hand.

If by 'intrinsic' you mean, however, that the feature is one that any object similar in construction specifiable in only local terms (lengths of parts, magnitudes of angles between parts and what we will soon call 'local handedness', for example) will have, irrespective of the nature of the space in which the object is embedded, then handedness is not intrinsic. For an object of a given construction so specified may not even be "handed" in the full sense at all — if it is in a nonorientable space, for example; whereas another possible object, describable in the same terms in the local way, but now taken as embedded in an orientable space, will indeed have full handedness. If 'intrinsic' means "independent of the topology of the space as a whole," then handedness is not intrinsic.

What does it mean, according to the relationist, for the hand to be a homomorph? Only that, given any possible counterpart to the hand, that is to say any possible object whose parts have the same lengths and the same magnitude of angles between them, then there is a possible crm that brings them into congruence.

Suppose that there are no enantiomorphs. In what sense can the hand still be said to be left-handed or right-handed? Well, suppose there is a possible crm that takes the hand into its counterpart because, although the hand is in an oriented three-space, the three-space is embedded in a four-space. Note that all this talk about "being in an oriented three-space" and "there being an embedding four-space" is all perfectly intelligible from a relationist standpoint. The assertions are, of course, explicated in terms of possible spatial relations among possible objects, and the lawlike features of the collection of such relations.

Now although any two three-dimensional hands which are counterparts and which are in the three-space will be bringable into congruence by a crm that takes at least one of the hands out of the three-space into other parts of the embedding four-space, it will still be the case that there are pairs of possible incongruous counterpart hands in the three-space, in the sense that (a) the members of the pair are counterparts, and (b) no crm that keeps both hands in the three-sub-space will bring the hands into congruence.

In this case, one of the hands will be "left in the three-space" and the other "right in the three-space." And each of the hands will be characterized by an "intrinsic feature": "being three-left" or "being three-right." The feature is intrinsic in the sense that an object that has it continues to have it no matter how many crm's it undergoes, *so long as the motions keep it confined to its original three-space.* What we see here is an illustration of the following general truth: What we mean by an intrinsic feature of an object is relative to some particular kind of transformation of the object we have in mind. A feature may be intrinsic relative to transformations of one kind, but not so relative to transformations of a different kind. There is nothing special about handedness here. For, as we saw, a structure can be intrinsically same-sided relative to one transformation (keeping the bars in the diagram in the plane) yet neither intrinsically same-sided nor intrinsically opposite-sided relative to some other transformation (allowing motions of the bars through all of three-space). That is just what we mean by 'intrinsic'.

Suppose that the hand is a homomorph despite the fact that it is in a three-space not contained in any embedding four-space. Once again, this statement is perfectly intelligible from a relationist point of view. Why is the hand a homomorph despite the absence of an embedding four-space? Perhaps because the three-space is globally nonorientable. Then all hands can be brought into congruence with their counterparts by global crm's; although there will still be possible pairs of hands such that (a) they are counterparts and (b) no *local* crm can bring them into congruence.

Suppose this is so. The hand is now obviously neither a left hand nor a right hand in the full sense. Kant thought this absurd, but Nerlich reasonably asserts that this just shows that Kant never considered the real possibility of higher-dimensional spaces or nonorientable spaces being the case. It is clear that our hand is now not "three-left" or "three-right" either. Is there any intrinsic feature of the hands which distinguishes the members of a locally incongruent pair of counterparts even though they are bringable into congruence by a global crm?

By this time the answer should be evident. In such a nonglobally-orientable three-space each hand is still either *locally-left* or *locally-right.* Of course there is no sense in asking whether a hand locally-left at point p is locally-left or locally-right when moved to point q. On the other hand, if p,q and the path taken by the object between them are all contained in a region of the space over which the local orientations can be extended in a globally consistent way, then partial global extensions of the purely local notion are possible.[4] Nor, of course, will it generally be true that an object locally-left at p at one time, then continuously rigidly moved about in the space and then returned to p will still be locally-left at p upon its return to the point of origin of its travels. That is just what it means to say that the space is globally nonorientable.

Notice also the following: Suppose we have a three-dimensional hand. Consider its "local-three-handedness." This will be an intrinsic property of the hand in the sense that it will be invariant under any local crm's that keep the hand confined to the original three-space in which it was located.

Even if that three-dimensional space is embedded in a four-dimensional superspace and even if the three-space is globally nonorientable, the local-three-handedness of the object is still well defined. It is, in fact, independent of the embedding of the three-space in any higher-dimensional space and independent of the global connectivity and orientability of the three-space. So in this sense of 'intrinsic', which connotes independence from questions of embedding of a space in higher-dimensional spaces and of global orientability of a space, local-three-handedness is a truly intrinsic feature of an object. I think that it is this fact that makes us want to say that a hand in our world is either left-handed or right-handed irrespective of the existence of any four-dimensional embedding space unknown to us or of the fact that the three-space of our world might, in fact, be globally nonorientable. The kind of handedness we normally have in mind is local-three-handedness, and this is independent of these possibilities about our physical space.

More fully and correctly: (a) we have very good reason to believe that there are locally-three-handed objects, since there clearly exist

4. In more detail the situation is like this: Even if a space is globally nonorientable there may be regions of it such that we can divide all the counterpart objects in the region into two classes of opposite handedness. An object of given handedness in the region cannot be brought into congruence with its counterpart of opposite handedness in the region by any crm which keeps the object in the region. If this is so, we can talk about "handedness with respect to the region." Of course in a nonorientable space the following situation can arise: (a) there is a region, A, such that there are pairs of objects in the region which are counterparts and such that no crm of the objects confined to A can bring them into congruence; (b) there is another region, B, partially overlapping A, which is, like A, regionally orientable; but (c) the region that is the union of A and B is such that any two counterparts in the united region can be brought into congruence by a crm in the united region.

possible counterparts incongruous under any local crm that keeps them confined to three-dimensional space; (b) we have very good reason to believe that this local-three-handedness can be extended to a regional-three-handedness over the presently observable spatial universe, since the three-space we are presently aware of seems quite globally orientable; (c) insofar as we have good reason to believe that our three-space is not embedded in a four-space, we have good reason to believe that there are regionally-handed objects; and (d) insofar as we have good reason to believe that our three-space is a globally orientable three-space, we have good reason to believe that there are handed objects in the world, *simpliciter*.

So all the crucial notions: being an enantiomorph, being a homomorph, having a specified number of dimensions, being orientable or nonorientable, being left and right—in the full, local, or "subspace" sense, are all completely intelligible from a relationist point of view.

(2) If someone objects to any of the notions I have invoked above, or, rather, to the particular definitions of particular geometric spaces or objects—such as enantiomorphism, local-handedness, etc.—let him ask himself whether postulating space as a substantival entity is going to leave him any better off.

If he wants to talk about the dimensionality of his "space" or its orientability, he will need just the same characterizations for this entity that we needed for our lawlike-governed collection of possible relations among possible material objects. Treating space as an object, rather than as a collection of possible relations among possible material objects, solves none of the difficulties in defining the various geometric notions necessary to characterize the "structure of space."

And if he objects to my postulation of relativized intrinsic features of objects, like enantiomorphism, local-enantiomorphism, or enantiomorphism-in-a-subspace, he should consider this: The only good that the postulation of substantival space is going to do for him in defining the relevant notions of the features of objects is that he will be able to predicate these features of the "space" in which an object is contained instead of predicating them of the object itself. He can then "explain" the nature of the object as being its relation to its containing space.

But this explanation seems no explanation at all, and the relationist objection is familiar. If all these features are well definable of the containing spaces of possible objects, why not just predicate them of the objects themselves and be done with it? The relationist argument, as always, consists partly in the claim that postulating substantival space provides no explanations and no understanding over and above postulating possible objects and their possible spatial relations. Worse yet, such postulation confuses the issue by making it look as though there were additional features of the world (for example, locations in substantival space) which really don't exist at all.

Notice that substantivalism is equally useless in answering such traditional positivist questions as: "How do we know which objects are left and which are right?" and "Mustn't there be some independent feature lawlike-connected with handedness in order for it to make sense to say that an object is left- or right-handed?" For if there are any real puzzles here, and I think that there are not but only confusions and pseudoproblems, then these real puzzles are just as much puzzles about handed pieces of space as they are about handed objects in space.

(3) What is the *real* substantivalist objection to relationism? The best philosophical argument I know is the claim either that (a) it makes no sense at all to talk about possible objects and possible spatial relations, but only about actual objects and their actual relations, and on this ground spatial talk is not wholly translatable into relationist talk without loss of content, or (b) that talk about possibilia makes sense only because of the underlying assumption of some actual substantival entity and its actual features, and in the case in question this can only be substantival space and its actual geometrical structure.

Now there are deep philosophical issues here. But one thing is clear, and that is the fundamental irrelevance of particular facts about enantiomorphism, homomorphism, or handedness. For if the relationist must invoke possibilia in order to explicate these notions, he must invoke them to explicate far more basic spatial concepts — for example, there being an empty spatial location in the actual world or there being a possible world of totally empty space. It was to account for these notions relationally that the idea of possible objects and their possible spatial relations was originally invoked. If the substantivalist wants to refute relationism on these grounds, he need not go to such recherché lengths as the invocation of questions about orientation and orientability. His quarrel with the relationist lies on much broader issues.

The only other arguments that I know for substantivalism which have any persuasiveness are those from particular aspects of physics, say from the "absoluteness" of absolute acceleration in Newtonian mechanics, special relativity, and, perhaps, general relativity as well. It is these arguments which, I take it, Earman finds persuasive. I am not sure that they are at all convincing, but in any case they hardly rest upon the possibility or nature of enantiomorphic, homomorphic, or handed objects.[5]

(4) Is handedness the result of a creative cause? Perhaps so. For there is neither more nor less reason to believe that the handedness of an object is the result of some causal factors than there is to believe any feature of the object to be the result of causes. But features of an object which consist in the relation of its parts to one another, or of the actual and possible relations of the object to other actual and possible objects, can surely be the result of causes. So even if handedness is "the result of the action of a creative cause," this in no way indicates anything inadequate whatever in the relationist account of what handedness *is*.

Is space "absolute?" Well, if that means "Is space a substance?" we have already seen the irrelevance of the consideration of notions of orientation to that issue. If it means "Do absolute motions in Newton's sense exist?" then surely nothing could be more irrelevant than the existence or nature of enantiomorphic objects.

Is space "primitive?" If that means that the existence of space temporally precedes there being objects in it, then all arguments from orientation are irrelevant. If it means that we can imagine space without material objects, but no spatial objects without space, then this is conceded by the relationist. There can be a collection of possible relations among possible objects without there being any objects or any actual relations among them; but

5. For a detailed discussion of the philosophical debate between the substantivalist and the relationist, and of the relevance or irrelevance of the results of physics to the philosophical debate, see L. Sklar, *Space, Time, and Spacetime* (Berkeley: University of California Press, 1974), chap. 3, "Absolute Motion and Substantival Spacetime."

there can't be an actual object with actual spatial features unless there are some possible objects with possible spatial features. Once again, questions of orientation are completely irrelevant.

If "Space is primitive" means "The relationist claim that all spatial talk can be translated into talk about the spatial relations among material objects is wrong," then, as we have seen, the allegation of primitiveness is just the allegation of the substantivalist about the incorrectness of the relationist account, and, as we have seen, the existence or the nature of features of orientation are completely irrelevant to the issue.

Is the relationist or the substantivalist account of space (or spacetime) correct? I don't know. What is clear, however, is that the existence and nature of incongruous counterparts are irrelevant to the issue.

If Kant and Nerlich are claiming that questions of orientation can be understood fully in the context of an understanding of the topology of space as a whole, then they are right. This is an important topological fact, and Kant may, indeed, have been anticipating this in his discussion of the handedness of objects. But if they are claiming that any facts about the existence or nature of orientation features add any weight to the other well-known arguments of the substantivalist against relationism, then they are plainly in error.[6]

6. Nerlich's replies to the arguments in this essay can be found in the context of his defense of the substantivalist position in his *The Shape of Space* (Cambridge: Cambridge University Press, 1976). See especially chap. 2, "Hands, Knees and Absolute Space." More on the role of incongruous counterparts in Kant's discovery and defense of his "transcendental idealist" account of space can be found in J. Buroker, *Space and Incongruence: The Origin of Kant's Idealism* (Dordrecht: Reidel, 1981).

9. What Might Be Right about the Causal Theory of Time

To what extent can we view an aspect of the spacetime structure of the world as in some sense reducible to or eliminable in terms of the causal structure of relations among events? This chapter takes up one special issue falling under the general question of the causal definability of spacetime structure. In particular I examine the plausibility of the claim that within the general relativistic context the full topological structure of spacetime can be causally defined.

Several different notions of causal structure are explicated. For some of these it turns out that the possibility of sufficiently causally pathological spacetimes precludes a causal definition of the topology in these senses of causal from being even extensionally adequate in all models compatible with the theory. A richer notion of causal structure is elucidated, and mathematical results about it explained, which does show that in a certain well-defined sense this new notion of causal structure is adequate in any general relativistic spacetime to fix the full spacetime topology. Even here, however, one must be cautious, as the topology is determined by this causal relation on the spacetime only if attention is restricted to a limited class of possible topologies, i.e., the usual manifold topologies.

Finally, I explore the alleged philosophical relevance of the mathematical results described. A notion of reduction which is epistemically motivated is described. From the point of view of this kind of reduction, the sense in which the full spacetime is reducible to a causal relationship among events seems not so much a reduction of the spatiotemporal to the causal as a reduction of the full spatiotemporal to that which is "immediately" epistemically accessible. While this provides a coherent rationale for the philosophical interest in such a notion of causal definability, it leaves us with many perplexing epistemic issues, which are then discussed.

The subject of this essay is also discussed in chapter 3, section VII, subsection "Causal Theories of Topology," and in chapter 10.

I

A CAUSAL theory of time, or, more properly, a causal theory of spacetime topology, might merely be the claim that according to some scientific theory (the true theory? our best confirmed theory to date?) some causal relationship among events is, as a matter of law or merely as a matter of physically contingent fact, coextensive with some relationship defined by the concepts of the topology of the spacetime. The strongest version of such a causal theory would be one which demonstrated such a coextensiveness between some causally definable notion and some concept of topology (such as "open set") sufficient to fully define all other topological notions. Given such a result, one would have demonstrated that for each topological aspect of the spacetime, a causal relationship among events could be found such that that causal relationship held when and only when the appropriate topological relationship held.

I believe, however, that the aim of those philosophers who have espoused causal theories of spacetime topology has been grander than this. There is a general and familiar philosophical program that works like this: In some given area of discourse it is alleged that our total epistemic access to the features of the world described in this discourse is exhausted by access to a set of entities, properties, and relations characterized by an apparently proper subset of this conceptual scheme. Then, it is alleged, the full content of propositions framed in this discourse must be characterizable in terms of its totality of observational content, this content being describable in the distinguished concepts of the subset of epistemically basic concepts. So, the program continues, it is up to us to show how all the empirical consequences of the theory really are framable in the distinguished concepts alone. Insofar as the theories framed in the totality of concepts appear to make empirical claims outrunning those expressible in the basic concepts, these must be shown to be ultimately reducible to the empirical content expressible in the basic concepts. And if the assertions of the total theory go beyond those reducible in content to assertions of the distinguished vocabulary, then they must be shown to be true merely as a matter of choice or "convention."

I believe that, explicitly or implicitly, some such program underlies all those various causal theories of spacetime topology with which we are familiar. So interpreted the philosophical theory is not just the explication of the results of some scientific theorizing, but an attempt at an epistemological and semantic critique which displays initial constraints into which any satisfactory scientific theory must fit. For if the epistemic and semantic analysis proposed by the philosophical theory is correct, then the limits of meaningful assertiveness of the scientific theories are delimited, the distinction between fact asserting and "mere convention making" elements in the scientific theories is drawn, and the criteria are made clear under which we are entitled to say whether or not two allegedly alternative total theories are or are not properly speaking "equivalent" to one another and, in Reichenbach's terms, whether they differ merely in descriptive simplicity or in some genuine empirical content.

Taking our total theory to be one which describes the spatiotemporal topological structure on events and that portion of the causal relationships among them which contains reference only to which events are and are not causally connectible with one another, a causal theory of spacetime topology is a philosophical theory which alleges that the total empirical content of this overall theory is exhausted by its causal part; and that any spatiotemporal topological assertion can be reduced to an assertion about causal relatability supplemented, perhaps, by conventional choices of "ways of speaking."

But, I believe, we must proceed with some caution here. What counts as "the full causal structure on events," and, hence, what counts as the allowable empirical basis to which all our theoretical elements are to be reduced may not be the same in all 'causal' theories of spacetime topology. And two philosophical theories even while agreeing as to the nature of the "epistemic basis," may, as we shall see, disagree as to *why* the elements in question are allowed into the reduction base.

What I wish to suggest here is that, for rather well-known reasons, some choices of an epistemological basis in causal theories of spacetime topology won't do the job intended for them. I think that one suggested basis does have some plausibility. But its plausibility, I will suggest, rests on grounds which might lead us to think that the designation 'causal' is inappropriate for this philosophical reductionist account of spacetime topology.

While our ultimate grounds for accepting or rejecting such a "philosophical" theory of spacetime topology will be "philosophical" or, more particularly, in terms of an epistemological critique of spacetime theories, it is important and enlightening to see just how particular scientific theories of spacetime fare in the light of one's philosophical account. I will try to show how several recent results in the mathematical foundations of general relativity "fit in" with the analysis of implausible and more plausible "causal" theories of topology which I will survey.

Finally, while I believe that the "causal" theory I end up with is a more plausible candidate for a possibly successful reductionist account of spacetime topology, I do not believe that it itself is uncontrovertibly correct. I will note some reasons for being cautious about accepting it as a philosophical account of spacetime epistemology and semantics.

II

Let us consider first the version of the theory which takes as its "reduction basis" the relation of causal connectibility among events. As usual, we note, the "definitions" of the topological notions in terms of the causal are in the mode of possibility. Some topological relationship is alleged to hold if some appropriate causal relationship *could* hold. That topological notions reduce to causal only if we allow causal notions "in the mode of possibility" is a familiar feature of such philosophical reductionism. I will forgo rehearsing the well-known reasons for the necessity of the invocation of such possibility talk and the equally familiar reasons for alleging that such an invocation vitiates at least part of the reductionist program.

In the version we are now exploring the notion of two events being causally connectible, i.e., its being possible for one to be causally related to the other, is taken as our primitive notion. In terms of it we attempt to construct an adequate defining basis for the full gamut of spatiotemporal topological notions. Not surprisingly, we do not need to invoke any notion of the direction of time at this point, and causal connectibility, rather than the asymmetric relationship of 'a could be a cause of b' will be sufficient.[1] How plausible is such a causal theory of topology?

1. R. Latzer, "Nondirected Light Signals and the Structure of Time," *Synthese* 24 (1972): 236–280. D. Malament, Ph.D. dissertation for Rockefeller University, "Some Problems Concerning the Causal Structure of Space-Time," chap. 2.

There are, I believe, two basic grounds on which a philosophical reductionist thesis is usually challenged:

(1) First, it might be claimed that in order to have grounds for believing that a relationship utilized in the reduction basis holds, we must first have some knowledge of relationships holding which are in the portion of the theory to be reduced or defined away. If this is so, it is claimed, then the reduction basis is not a properly epistemically primitive basis.

(2) Second, it can be argued that the reduction basis is not adequate to provide "definitions" for all the concepts of the total theory. That is, even if we can, without knowing any of the defined relationships to hold, establish which relationships in the reduction basis do in fact hold, that information will not be adequate to fully determine all the relationships which are expressed in the total theory.

Our first version of a causal theory of spacetime topology has been challenged on both these grounds.

First, can we really tell which events are causally connectible without already knowing a great deal about their spatiotemporal relationships to one another? In our usual accounts of the notion of causality we find that while causality means falling under a natural law, and a law is a general rule connecting *kinds* of events, knowing which *particular* events are causally related to which other particular events requires knowing a great deal about the spatiotemporal relations of these events to one another. Strikings of matches cause lightings of matches, but to tell that *this* striking caused *that* lighting we must know that this particular striking bears to that particular lighting an appropriate spatiotemporal relationship. Now while I believe that there is some truth to this objection, I think that a possibly viable "causal" theory of topology is lurking under this first theory despite this initial objection. But rather than pursue the matter here, I will save this discussion for later.[2]

Second, can we really define all our topological notions in terms of the notion of causal connectibility? Here is where some results of the mathematical study of spacetimes become crucial. If the philosophical reductionist theory is correct, it is claimed, then in any

2. For a statement of this kind of objection to causal theories as epistemologically motivated reductions, see A. Grünbaum, *Philosophical Problems of Space and Time* (New York: Knopf, 1963), 190–191.

possible spacetime we envision the topological notions must be connected to the appropriate causal notions in the way specified by the reductionist account. For the topological notions are supposed to be *defined* by the causal notions. Is there a definition of the appropriate basis for all topological notions (say a definition of 'open set') in terms of causal connectibility which holds in all the spacetimes we take as conceivable?

Not if we allow ourselves the full range of spacetime compatible with general relativity, including those we might view as causally "pathological."

In the Minkowski spacetime of special relativity we can indeed causally define (in the present meaning of that term) the open set basis sufficient to fully define the topology. An explicit definition of the open sets in terms of causal connectibility is available in terms of the well-known Alexandroff topology for Minkowski spacetime.

In those general relativistic spacetimes which are *strongly-causal* it can also be shown that the Alexandroff topology and the usual manifold topology will coincide. Indeed, the coincidence of the topology defined by the causally defined open sets of the Alexandroff topology with the usual manifold topology is equivalent to strong causality. When the spacetime is not strongly-causal, however, the manifold topology and the Alexandroff topology will fail to coincide.[3]

Malament has recently shown that in any spacetime which is both *past* and *future distinguishing* (a weaker condition than strong causality) one can at least *implicitly* define the topology in terms of causal connectibility. That is, given any two past and future distinguishing spacetimes, any causal isomorphism between them (bijection preserving causal connectibility) will be a homeomorphism relative to the usual manifold topology. He has also shown that this is a strongest possible result in that one can construct examples of spacetime in which either past or future distinguishing is violated and such that there will be causal isomorphisms between pairs of such spacetimes which fail to be homeomorphisms. If the past and future distinguishing condition is not met the topology cannot be

3. See S. Hawking and G. Ellis, *The Large Scale Structure of Space-Time* (Cambridge: Cambridge University Press, 1973), 196–197; and S. Hawking, A. King, and P. McCarthy, "A New Topology for Curved Space-Time Which Incorporates the Causal, Differential, and Conformal Structures," *Journal of Mathematical Physics* 17 (1976): 174–181, especially 176.

even implicitly defined in terms of causal connectibility. There are, in fact, spacetimes where the failure of causal definition of the topology is made particularly manifest. I refer here to those space-times where the topology is nontrivial but where causal connecti-bility is a relationship which holds between *every* pair of events.[4]

Now we might try to hold that the kinds of spacetimes in which either explicit or implicit definition of the topology by causal con-nectibility fails are impossible spacetimes. For example, they are, to be sure, infected with the kind of pathological causality which we would not ordinarily expect to find in a causally "well behaved" universe. But if we can even *understand* what such spacetimes would be like, then the fact that we can understand them (know what it would be like to live in one of them, for example) shows us, I think, that any hopes of establishing that in the actual world our actual concepts of topology are *defined* by notions of causal con-nectibility are unacceptable. Without going into details here, I be-lieve that such worlds are perfectly intelligible to us and that with a little imagination anyone familiar with the basic concepts of rela-tivistic spacetime can be gotten to understand just what the topo-logical aspects of such a world would be like to an inhabitant of them. Further, we can construct pairs of spacetimes which are causally isomorphic in our present sense and which are such that an inhabitant of them can, by topological exploration, discover in which of the alternative spacetimes he resides.[5]

III

Now the fact that our initial attempt at a causal theory of spacetime topology fails in two diametrically opposite directions might lead us to believe that the prospects for any such theory are dismal indeed. The basis chosen in this original version of the theory is too *weak* to do the job required of it — define the full topological struc-ture in every spacetime we wish to take as intelligible. Yet it is so *strong,* i.e., contains so many elements and of such a kind, that it

4. See chap. 3 in Malament's thesis cited in n. 1 above, and his "The Class of Continuous Timelike Curves Determines the Topology of Spacetime," *Journal of Mathematical Physics* 18 (1977): 1399–1404, especially p. 1401, Theorem 2.
5. To see this, consider the example used by Malament on p. 1402 of his "The Class of Continuous Timelike Curves" to show that a failure of past or future distinguishability is sufficient to generate causal isomorphisms which are not homeomorphisms.

appears that we already need epistemic access to much spatiotemporal structure on events in order to tell when the relationship
which is utilized in the basis, causal connectibility, holds. But we
will see that the situation is not quite as hopeless as it looks.

Let us try to remedy the defects of our original theory first by
moving to a stronger basis, holding in abeyance for the moment the
obvious difficulties this will give rise to. Within the relativistic
context two events are causally connectible if there is a continuous
causal (timelike or lightlike) path between them. This follows from
the standard view that causal interaction is inevitably mediated by
the emission, transmission and reception of some "genidentical"
material or lightlike "particle." So within this context, we know
that the relationship of causal connectibility holds of a pair of
events if we know the truth of the existential claim that there is at
least one continuous causal path containing the two events.

Now suppose we take as our basis notion not causal connectibility but, instead, the notion of continuous causal path. We then have
full knowledge of the basis relationship not when we know merely
which pairs of events are connectible by some continuous causal
path or other, but when we can tell of any set of events whether or
not it constitutes a continuous causal path. Does the introduction
of this much stronger basis change the picture significantly?

That it does follows from an important and interesting result of
Malament's. Given any two spacetimes (taken to be four-dimensional manifolds with pseudo-Riemannian metrics of Lorentzian
signature) any bijection which takes continuous causal curves into
continuous causal curves will be a homeomorphism! In other
words, the full topology of the spacetime will be fixed by its class of
continuous causal curves. So if we can fully determine the latter,
then we can fully specify the former as well.[6]

But can we determine the continuous causal curves of a spacetime without already being able to determine its topological features? This obvious objection to the epistemological use of a result
like Malament's to establish a causal theory of topology is well
known. Let us first present it and then in the next section move on
to the deferred task of showing why, despite this argument, something which might be called a causal theory of topology can have a
case made out for it that meets this "epistemological" objection.

6. Malament, "Some Problems," chap. 3, and "The Class of Continuous Timelike Curves," p. 1401, Theorem 1.

In one version, and among philosophers I think the most common one, the picture one has is something like this: A single observer wishes to map out the topology of the spacetime in which he dwells. Equipping himself with an infinite number of material or lightlike particles, he emits them from all points in all directions (we must be generous in such idealizations) so determining the structure of continuous causal paths in his spacetime. Now, by Malament's result, he can pin down uniquely the full topology of the spacetime, including such not-directly-determined features as which spacelike paths are continuous, etc.

But how does our observer tell which classes of events in the life history of a "genidentical" particle are spatiotemporally continuous portions of its history? Indeed, how does he even tell which events are events in the history of one such particle and not events selected at random from the histories of any number of distinct particles? Only, it is alleged, by already knowing what the continuous spatiotemporal segments really are, i.e., only by already having access to topological features of the spacetime.[7]

IV

But consider the following version of a "causal" theory of topology: It is true that on the picture just looked at we gain access to the full topology of the spacetime only by already knowing part of it, i.e., only by already knowing which are the continuous timelike and lightlike curves in it. But this is already a "reductionist" gain. And when we realize the real epistemological importance of the basis to which total topological knowledge has been reduced, we see just how important a gain this is. For what constitutes the continuous causal paths of the spacetime is just what is available to our *direct epistemic access*. Not because we have some wonderful way of spotting continuous segments in the history of genidentical particles without any antecedent knowledge of the topology. That is silly, for what we mean by continuous segment in the history of a genidentical particle is just a continuous set of spacetime locations along a causal path occupied by the same kind of material particle.

7. For this objection, see H. Lacey, "The Causal Theory of Time: A Critique of Grünbaum's Version," *Philosophy of Science* 35 (1968): 332–354. See also L. Sklar, *Space, Time, and Spacetime* (Berkeley: University of California Press, 1974), sec. IV, E, 3.

But, rather, because each causal path can be traversed by a local "consciousness" who directly and immediately, as a primitive content of his experience, can tell which events in his consciousness, and hence which spacetime locations along the worldline of his history, are "near" one another!

The idealized picture we have now is of a universe equipped not with one observing "consciousness" but with a plentitude of them, so that the totality of causal worldlines of the spacetime is covered by consciousnesses sensing immediately and directly which locations of the spacetime along their respective worldlines are "near," i.e., determining without inference or instrument which are the continuous segments of causal curves in the spacetime. Extravagant as this seems when put this way, the basic idea here has appeared explicitly or implicitly from Robb through Hawking and Malament.[8]

So the causal theory of spacetime topology has now been replaced by a reductionist account which is not really "causal" in its fundamental epistemic motivation. The reductionist program still takes as the body of concepts to be reduced those characterizing the full topology of the spacetime. The reduction basis consists in those concepts relating to continuity of the one-dimensional causal worldlines of the spacetime. But this set of topological features is not discriminated from the general class because the causal worldlines are the paths of propagation of causal influence throughout the spacetime but, rather, because such curves, being the worldlines of the possible life histories of "consciousnesses," have their continuity open to "direct epistemic access" without the intervention of instruments or the necessity of theoretical inference. I will shortly return to the question of just how plausible such a reductionist program might be. I don't wish here to maintain that this is the correct philosophical account of our knowledge of the topology of

8. See the introductory portions of A. Robb, *A Theory of Time and Space* (Cambridge: Cambridge University Press, 1914), where the epistemic suitability of taking 'after' as the basic primitive for constructing a spacetime theory is discussed. See Hawking, King, and McCarthy, "A New Topology," 175. In "The Class of Continuous Timelike Curves" Malament on p. 1399 says that his result on the implicit definability of the topology by the class of continuous timelike curves is "of interest because, in at least some sense, we directly experience whether events on our worldlines are 'close' or not."

spacetime and of the semantic analysis of topological concepts in general, but only that it is the most plausible version of a "causal" theory of topology, combining, as it does, a genuine reduction (from the totality of topological aspects to those along causal paths alone) with an at least prima facie case for the epistemic priority (in terms of immediacy of access) of the concepts of the reduction basis. I also believe that at least some readers will agree with me that it is this version of the "causal" theories which captures the fundamental epistemic "intuitions" which lay behind the other more "causal" causal accounts.

V

At this point it is enlightening to introduce some additional results of recent mathematical work on spacetimes. At first glance these new results might seem to militate against the version of the causal theory we ended up with. But on reflection, I believe, they illustrate, rather, the force of our last account of the causal theory.

Suppose we have available to us a full picture of the topology (continuity) of the causal worldlines of the spacetime. Can we then infer immediately the full topology of the spacetime? Malament's result seems to show that we can, for any two spacetimes (in the usual 4-manifold with Lorentz-signature metric sense) which agree on their causal worldline topology will agree in all topological aspects. But it isn't that simple.

Following the work of Zeeman on novel topologies for Minkowski spacetime, recent work has been done on looking for interesting novel topologies for the spacetime of general relativistic worlds. These new topologies, unlike the usual manifold topology, "code" into themselves the full causal structure of the spacetime. This is true in the sense that any homeomorphism between spacetimes relative to the new topologies are automatically conformal isometries. This is not true of homeomorphism relative to the usual manifold topologies. With these new topologies the spacetimes are not manifolds.[9]

9. See E. Zeeman, "The Topology of Minkowski Spacetime," *Topology* 6 (1967): 161–170, for the introduction of these nonmanifold topologies in special relativity. For the generalization to general relativistic spacetimes, see Hawking, King, and McCarthy, "A New Topology."

The importance of these new topologies for our present purposes can be seen in the manner of their construction. One starts off imagining a spacetime equipped with the usual manifold topology. One then seeks a new topology which will code the causal structure in the manner described above, and which will then, perforce, differ from the usual manifold topology. But the new topology is so constructed that it *agrees with the manifold topology on the topology induced on the one-dimensional causal curves.* For example, there is the *path topology* of Hawking, King, and McCarthy. It is defined as being the finest topology on the spacetime which induces on all continuous causal curves the same topology induced on them by the usual manifold topology.

So, obviously, the standard and the novel topologies will agree on all topological facts about the causal curves. Now aren't we in a position to be skeptical about our "causal theory of topology"? For if we can fully exhaust our knowledge about the topology along such causal paths and yet still not know the full topology of the spacetime (is it the standard manifold topology or is it, instead, one of the novel topologies?) aren't we in just the same position which caused us to reject our first version of a causal theory of topology? For there we pointed out that in the case of spacetimes which were not both past and future distinguishing, two manifolds could share their causal structure (there could be a causal isomorphism between them) and yet not be topologically identical (i.e., not homeomorphic). This led us to say that the structure determined by causal connectibility was not sufficient, in general, to fully fix the topology of the spacetime. And here, once again, we seem to be saying that once we allow for the possibility of the novel topologies, we can no longer fully fix the topology of the spacetime even given full topological knowledge of the causal paths.

But the situations are not parallel. The trouble with the causal connectibility version of the causal theory of topology is this: Two spacetimes might be causally isomorphic and yet be *empirically distinguishable* in their topological structure. They may very well differ in the structure of continuity along causal paths which is the paradigm of epistemically accessible topological structure. But if two spacetimes share the same topological structure along the causal paths, and if this structure exhausts the empirically determinable topological structure of spacetime, then no empirical observation could tell us which of two incompatible full topologies (say standard vs. the path topology of Hawking et al.) is *really* the full topology.

And what that suggests is this (the familiar end product of the problem of theoretical conventionality): If two spacetimes share the same topology induced on the causal paths, then, appearances to the contrary, they are really (topologically) the same spacetime. Their real topological structure may be *expressed* in the standard manifold form (and, as Malament's theorem shows, only one of these will be compatible with the topology on the causal paths) or it may be expressed in terms of the path topology of Hawking et al., or in terms of any other nonmanifold topology designed to code the causal structure. But these are merely alternative formulations of one and the same set of empirical facts. For (in the usual positivist vein) since the topology on the paths exhausts what can be empirically discovered about the topology in general, the total factual content of the general topology is exhausted by the topological structure it induces on the causal paths.

One awaits at the present time a neat mathematical formulation which eschews the necessity for topological assertions about the spacetime which *appear* to outrun the topological facts about causal curves, and which captures the empirical equivalence among all topological structures which induce the same topology on the causal paths, by offering as the full characterization of the topology of the spacetime the mathematical description of the continuity structure on causal paths alone.

VI

How plausible is a causal theory of spacetime topology, even in its most plausible version in which it becomes rather a "topology of world-lines traversable by a consciousness" theory of spacetime topology? Now, of course, much of the answer to that question will depend upon just how plausible in general one takes philosophical reductionist theses, with their underlying positivist motivations, to be. While many are skeptical (often without much in the way of argument above and beyond "realist" dogmatism) of reductionism in general, this is hardly the place to rehearse most of the familiar issues of whether or not "pure" observation bases exist, of the allowability of modal possibility talk into the reductive definitions, etc. Rather let me focus on just a few issues for discussion which come to mind when one reflects on this particular attempt at philosophical reductionism.[10]

(1) The theory presupposes that at least one aspect of spacetime structure is accessible to consciousness without instrument or inference. But is this belief supportable? Here the reader may remember the discussion of spatiotemporal coincidence in Reichenbach's *Axiomatization of the Theory of Relativity*. Here Reichenbach reflects upon the fact that while coincidence is taken as a primitive of his (and everyone else's) reconstruction of relativity, if this decision to take it as a primitive is based upon the ground that we have immediate noninferential knowledge of what events in the physical spatiotemporal world are really coincident we are on dangerous ground. Aren't the coincidences of which we are "immediately" aware *subjective* coincidences; and isn't it *objective* coincidence with which we are concerned in our reconstruction of the theory of physical spacetime?[11]

In the present case, by what right do we identify the continuity of inner experience of some observer, as subjectively experienced, with the "real" physical continuity of the spacetime locations at which these experiences occur? Imaging, for example, a consciousness instantaneously and discontinuously transported through spacetime, so that the continuity of his "inner" experiences *misrepresents* the discontinuity of the spatiotemporal locations at which he has them.

What we have here is just one more example of a well-known and enormous difficulty for reductionism: If reductionism allows in its basis for reconstruction only the content of the immediately experienced, then how can it ever give us an adequate account of the nature of the objective, physical world? For if we slide down the slippery slope we get on to when we once begin to "reduce" the inferred to the "immediately experienced," how can we ever stop before reaching the bottom where the basis is the contents of sub-

10. For a discussion of reductionism in general and in the specific context of spacetime theories, see Sklar, *Space, Time, and Spacetime,* sec. II, H, 4, and "Fact, Conventions, and Assumptions in the Theory of Spacetime," chap. 3 in this volume.

11. See H. Reichenbach, *Axiomatization of the Theory of Relativity* (Berkeley: University of California Press, 1969), 16–21. See also chap. 3 in this volume, sec. IV, subsection "A Formalization Differing in Definitional Consequence Only."

jective experience; thereby dropping out of the realm of concepts dealing with the physical "outer" world altogether?

(2) The theory presupposes that our *only* direct epistemic access is to the continuity of spacetime locations in the lived one-dimensional history of a consciousness traversing a causal worldline. Is this plausible?

First, relying on something quite reminiscent of Kant's distinction between space as the manifold of apperception of outer experience and time as the manifold of apperception of both outer and inner experience, the theory places our experience of time, and continuity in it, in a very special position indeed. All our topological insight into the world is to be grounded in an awareness only of the *temporal* continuity of our experience. Our knowledge of the spatial aspects of the world is derivative from this temporal sense, and, if what we have said about conventionally alternative topologies is correct, really, properly speaking, only conventional. For the only *real* topological facts are the facts about temporal continuity along causal paths.

But is this correct? What about our apparent "direct" knowledge of the structure of the space around us? Is this to be dismissed? We could, of course, pull the not uncommon move of distinguishing "perceptual space" from "physical space" and argue that our direct apprehensions are only of the former, the latter to be "constructed" out of our temporal experience. But then the peculiar asymmetry of taking most of our perceptual life to be "merely subjective" while allowing our direct experience of temporal continuity to serve as immediate access to the external world appears in an even more striking light.

Consider, again, what is, on reflection, the rather surprisingly different treatment accorded topological and metrical aspects of spacetime on this view. Nearly everyone who writes about the foundations of our knowledge of spacetime takes metric features of the world, even those along causal paths (elapsed proper time along a causal world-line) to be founded on some physical measuring process. In order to ground our knowledge of just how much proper time elapses between two events on a causal path we must rely, it is almost invariably alleged, either on *clocks* or on a specification of the affine parameter determined by the paths of *freely falling particles.* If one tries to bring forward our immediate experience of the magnitude of duration, this is almost always dismissed as a confusion of "psychological time" with the objective physical magnitude.

But why should psychological experience of temporal magnitude be irrelevant to founding our knowledge of the magnitude of physical duration and yet the psychological experience of temporal continuity be taken as giving us immediate access to the real facts about continuity along causal paths?

(3) Next consider Reichenbach's well-known claim that the topology of spacetime is just as "conventional" as is its metric. Just as we can save alternative metrics in the light of any experience by a sufficiently rich flexibility in the postulation of universal forces, so any topological thesis can be maintained if we allow ourselves the global flexibility of choosing how to identify or dis-identify events (moving from multiply connected spaces to their simply connected covering spaces, for example) and the local flexibility of tolerating causal anomalies, i.e., spatiotemporally discontinuous causal interaction mediated by the spatiotemporally discontinuous motion of genidentical signals.[12]

But if the continuity of world-lines is determinable directly by consciousness traversing them, aren't we denied at least the second kind of flexibility noted? So aren't at least the local facts about topology nonconventional? Here I think that two (obvious) options present themselves: (a) We could accept the nonconventionality of continuity along causal world-lines, thereby supporting the intuitions which at least some have had that topological facts (or at least some of them) are nonconventional in opposition to the full conventionality of the metric; or (b) We could remember the possibility, noted above, of consciousnesses themselves experiencing as continuous what are actually spatiotemporally discontinuous histories, and allow for the saving of topologies now by contemplation of a sufficiently rich allowance of "experientially anomalous" worlds.

All of this is just one more way of emphasizing the fundamentally problematic nature of our best causal theory: the fact that it seems to put one kind of experience on our part — the experience of temporal continuity — on a pedestal as the *one* way in which psy-

---

12. See H. Reichenbach, *The Philosophy of Space and Time* (New York: Dover, 1958), 65.

chological experience gives us "direct access" to the structure of the objective world.

(4) It is worthwhile to reflect for a moment on the idealized basis of "directly knowable facts" to which all assertions are to be reduced in our theory of topology. This consists in all facts about the continuity of causal path segments, for each such segment could be traversed by a consciousness able to directly ascertain its topology.

But, of course, no one consciousness, even in the most extreme idealization, could traverse them all. For example, a consciousness which experiences one event is, in principle, excluded from having in its experience, ever, the direct awareness of an event at spacelike separation from this event.

Of course it is a standard problem with philosophical reductionisms of this kind as to whether the reduction basis should consist in the total possible experience of a *single* observer or the total amalgamated experience of *all possible* observers. In the relativistic context this takes on a particularly disturbing aspect. The experiences of some observers simply can't be communicated to some other observers. It is easy to cook up spacetimes such that one has a pair of them which are distinct in their topology and metric structure; whose differences are determinable on the basis of the collective experience of all possible observers; but which are such that no one observer could ever tell — even if he lived a world-line of infinite extent past and future — which of the two spacetimes he lived in. These are the so-called *indistinguishable* spacetimes.[13]

Should a reductionist account of topology allow in its basis of facts everything which could be known to all observers taken collectively, or should the basis consist, rather, in the possible experience of some one observer? The former alternative seems in many ways the more natural, for even on radically positivist grounds it seems unfair to exclude as "real facts about the world" something which *someone* could ascertain in a direct noninferential way. On

13. On indistinguishable spacetimes, see C. Glymour, "Topology, Cosmology and Convention," *Synthese* 24 (1972): 195–218. See also Glymour's "Indistinguishable Space-Times and the Fundamental Group," and D. Malament's "Observationally Indistinguishable Space-Times," both in J. Earman, C. Glymour, and J. Stachel, eds., *Foundations of Space-Time Theories, Minnesota Studies in the Philosophy of Science,* vol. 8 (Minnesota: University of Minnesota Press, 1977).

the other hand, the usual pressures toward solipsism encountered in positivist programs may tend to make a sufficiently radical reductionist reject even the basis to which the causal theory reduces topology as allowing too much in, since it takes as factual elements of the world which could never, even in the most "in principle" way, be known to any single observer trying to ascertain the topology of the world.

VII

Ultimately, deciding on the plausibility of some reductionist account of topology of the kind we have been exploring will require resolving some of the deepest of philosophical questions: questions relating to the existence and nature of an "observation basis" on which all theorization is to be constructed and questions relating to just what extent one can, on the one hand, support a positivist-reductionist account of "theoretical" features of the world and, on the other hand, to what extent one can defend a "realism" which allows for valid inference "beyond the immediately observable." A theory of theories which does justice both to our strongly held intuitions of realism and yet at the same time makes coherent sense of the progress of physics in weeding out "metaphysical" elements by means of epistemic critiques and Ockham's Razorish prunings of theories is not yet before us. But at least I think that we can now see that a causal theory of topology is really one more attempt at such an epistemological critique of theories.

I have been maintaining that 'causal' is really a misnomer for such theories of topology. To be sure even in our last, most plausible, version of a "causal" theory of topology the "hard facts" about the topology of the world are reduced to facts about the continuity and discontinuity of causal paths in the spacetime. But not *because* they are the paths of "genidentical" causal signals, rather because they are the paths which constitute the possible life-histories of experiencing consciousnesses.

One final query: Is the identity of the causal paths with the world-lines of possible consciousnesses just an "accident," just an artifact, as it were, of the relativistic facts about the lawlike nature of the world? Or is this identity within the relativistic context instead the inevitable result of any physical theory which survives an adequate epistemic critique? If the latter is really the situation, rather than the former, then the misidentification of reductionist topological theories as causal may not be merely a mistake, but rather a symptom leading us to further insight.[14]

14. Consider a possible world inhabited by "epiphenomenal" consciousnesses able to experience events at their locations but not capable of exerting causal influences. Now imagine causal signals to be confined to a subclass of paths in the world, as in ours, but these consciousnesses able to traverse what are spacelike paths. Would not the "epistemic basis" for our spacetime topology in such a world outrun the continuity of causal paths in that it would include the directly experienceable continuity of paths traversable by a consciousness but not by a causal signal? Is such a world imaginable? Why not? On the other hand, one is, at least at first, more skeptical of the possibility of a world where the causal paths outrun those traversable by a consciousness, for just how would we "ultimately" establish the genidentity of the signals marking out those causal paths not open to the direct topological inspection of some idealized local consciousness? Would, if such a world were possible, our epistemic basis for the topology once again be limited to the (now narrower) class of "directly experienceable" paths?

10. Prospects for a Causal Theory of Spacetime

This chapter attempts to provide a general framework for discussing the plausibility of various so-called causal theories of spacetime. Two distinct classes of such theories are delimited. In the first class are those attempts at reducing spatiotemporal to causal structure which rely upon an allegation that our full epistemic access into spacetime structure is through direct epistemic access to a limited class of relationships among events. The second class of theories are those which allege that spatiotemporal structure is reducible to causal structure in a manner more like that in which one level of entities and properties in the world is shown to be identical to some other level of entities and properties by means of a scientific reduction proceeding through an identification. In both cases pushing the allegation of causal reduction to its limits results in philosophical perplexity.

Examples of the first kind of reduction are the so-called causal theories of the metric of Minkowski spacetime (as in Robb), and causal theories of spacetime topology in general relativity. In these cases there is first an initial problem of guaranteeing that the causal structure invoked is sufficient to extensionally fix the spacetime structure in question in any model of the world allowed by the theory in question. Even after this problem has been resolved, there are deep philosophical difficulties in justifying the epistemic foundational claims needed to motivate the reductionist program. Finally, there is the tendency of such programs, endemic to epistemically motivated reductionist programs, to push us into unacceptably solipsistic views of the world.

Examples of the identificatory reductive programs are then sketched. Here again there are certain fundamental philosophical obstacles to rationalizing the reductions. Finally, pushing the claims of such identificatory reductions to their limits tends, in a familiar manner, to separate off the world of experience from the world hypothesized by our natural science, in a manner which may leave us perplexed when such fundamental features of the world as its spatiotemporality, as we experience it, are reduced to mere "secondary qualities."

Some of the material discussed here has been treated at greater length in chapter 3 and in chapter 9.

W HAT could possibly constitute a more essential, a more ineliminable, component of our conceptual framework than that ordering of phenomena which places them in space and time? The spatiality and temporality of things is, we feel, the very condition of their existing at all and having other, less primordial, features. A world devoid of color, smell, or taste we could, perhaps, imagine. Similarly, a world stripped of what we take to be essential theoretical properties also seems conceivable to us. We could imagine a world without electric charge, without the atomic constitution of matter, perhaps without matter at all. But a world not in time? A world not spatial? Except to some Platonists, I suppose, such a world seems devoid of real being altogether.

Given this sense of the primordialness and ineliminability of spatiotemporality, it is not surprising that many have tried, in one way or another, to account for or explain away other features of the world in terms of spatiotemporal features. Witness Descartes on extension as the sole real mode of matter and the Newtonian-Lockean program in general.

Yet there is a countercurrent to this one. There is a collection of programs motivated by the underlying idea that either all spatiotemporal features of the world are to be reduced to other features not prima facie spatiotemporal, or at least that some of the spatiotemporal features of the world are to be eliminated from a fundamental account in terms of a subset of themselves and other features not prima facie spatiotemporal. Collectively these programs are sometimes said to espouse "causal" theories of spacetime. As we shall see, there is a wide range of such programs with a variety of conclusions as their goals and with quite different underlying rationales. Some, indeed, probably ought not to be called causal theories at all.

What I would like to do here is to examine at least two classes of such programs, explore some examples within these classes, disentangle some of the confusions which have resulted from an insufficient awareness that such a diversity of goals and presuppositions exists in so-called causal theories of spacetime, and, finally, to suggest that for all the diversity both programs lead in the end to quite familiar and related kinds of philosophical perplexity.

I

Both classes of causal theory I will discuss have as their members theories which allege that some or all spatiotemporal relations "reduce to" or are "eliminable in terms of" some class of relations, relations either not spatiotemporal at all or, at least, members of a modest subset of the original presumed totality of primitive spatiotemporal relations. In the theories of both classes it is alleged that the connection between associated "reduced" relation and "reducing" relation is more than a mere accidental coextensiveness. Indeed, more than lawlike coextensiveness will be claimed for the relations. In both cases the association will be declared necessary, although, as we shall see, the grounds for the allegation of necessity of the association (if not the kind of necessity) will be very different in the two cases.

But the motivation for alleging that the reduction ought to be carried out will differ markedly in the two cases. And the kinds of arguments which can be brought forward to either support the claim of reduction or attack it will differ as well.

What are the two classes of "causal" theories of spacetime? Crudely, the distinction is this: some "causal" theories tell us that a spatiotemporal relation must be considered as reduced to a causal relation because our full and complete epistemic access to the spatiotemporal relation is by means of the causal relation, and therefore, some variant of a verificationist theory of meaning tells us that the very meaning of the predicate specifying the spatiotemporal relation is to be explicated in terms of the predicates expressing the causal relation. The other "causal" theories tell us that we are to take the spatiotemporal relation to be reduced to the causal relation because it is a scientific discovery that the former relation is identical to the latter.

Let us look at some features of the epistemically motivated alleged reductions first.

II

The idea that spatiotemporal relations must be identified with causal relations on epistemic grounds has its origins both in general empiricist and positivist claims about the association of meaning with mode of verification and, of course, in Einstein's critique of the notion of simultaneity for distant events. The basic line is that, just as Einstein showed us that we could only make concrete sense of the notion of simultaneity by identifying that relation with some specific causal relation among events, so in general a concept is legitimate if and only if a physically possible mode of verification is associated with its application. This shows us that in general nonlocal spatiotemporal notions can only be understood by their identification with some appropriate causal relation among events.

There are a number of common features to causal theories of spacetime of this sort which it would be well to point out.

First, there is a distinction drawn between those spatiotemporal features of the world to which epistemic access is taken as primitive. These relations remain in the reduction basis to which other spatiotemporal relations are reduced. Generally it is *local* spatiotemporal relations which have this unquestioned status. For example, in the Einstein critique of distant simultaneity, simultaneity for events at a point is never in question. It is simply assumed that such a relation exists and can be known to hold or not hold without the invocation of some causal means of verification. In this, and in other contexts as we shall see, the notion of what constitutes a continuous causal path in spacetime is also taken as primitive. In the Einstein "definition" for distant simultaneity this appears in the implicit assumption that we can identify one and the self-same light signal throughout its history.

For spatiotemporal relations which don't fall into the primitive class, whatever that is taken to be, it is assumed that one must construct a meaning for such relations out of those available. Here it is frequently alleged, implicitly, that a distinction must be drawn between two sorts of defined spatiotemporal relations—those which have genuine fact-like status and those which merely attribute relations to the world reflective of conventions or arbitrary decisions on our part. For example, in the special relativistic case it is frequently alleged that local simultaneity and causal connectibility are facts about events in the world, but distant simultaneity is merely a matter of convention. Such arguments are usually backed up by claims to the effect that which events are simultaneous at a point and which causally connectible is not a matter of theoretical choice on our part, but that we could redescribe the world in such a way as to change the distant simultaneity relations among events leaving the genuinely observational predictions of our new theory unchanged from those of an older account which ascribed different simultaneity relations for distant events.

Built into these epistemically motivated reductions is an implicit distinction between a priori and empirically given epistemic limitations. In the Einstein critique, for example, the assumption that at least local simultaneity is directly given epistemically, and hence not up for epistemic critique, is one which is simply imported into the critical situation in an unquestioned way. What the ultimate roots of this assumption are remains in question. All that I am noting here is that within the context of this particular epistemically motivated critique, that this particular relation is "given" to us simply is taken for granted a priori.

On the other hand, it is taken as an empirical matter, very much of interest in the context of the particular critique, as to just what the nature of the causal relationships are which allow us to extend outward from our primitive base of local spatiotemporal relations to a full set of spatiotemporal relations by means of causal definitions. For example, it is crucial to the Einstein critique that light be the maximally fast causal signal, that transported clocks not provide a unique synchronization, etc. Essentially, concepts are going to be admitted only if there can be associated with them a means of verification and, using the old positivist terminology, the limits of verification are going to be taken to be what is physically possibly verifiable, making the question of just what physical processes are lawlike allowed crucial for understanding our limits of legitimate concept introduction by definition.

Epistemically motivated causal theories of spacetime will generally attribute to some associations of spatiotemporal and causal relations a necessary status. Here the notion of necessity is familiar. If a spatiotemporal relation not in the primitive basis is to be introduced into our theory by means of a definitional association with some causal relation, then the meaning of the spatiotemporal relation is fixed by its definition and the connection of the spatiotemporal relation and the causal is one of analyticity and, hence, of necessity. We will have more to say of this shortly.

Finally, it is interesting to note that in many such epistemically motivated causal theories of spacetime, the real set of primitive relations is plausibly taken to be itself a set of spatiotemporal relations. In the relativistic critique, for example, local simultaneity is plainly itself a spatiotemporal relation. But what about the all-important notion of causal connectibility? When one realizes that the critique takes causal connectibility to be connectibility by a geni-dentical signal of the velocity of light or less, and that, at least plausibly, genidentity seems to be at least in part, and in crucial part, spatiotemporal continuity, one begins to see the motivation behind the claim that such "causal" theories of spacetime might better be construed as epistemically motivated attempts to reduce the totality of spatiotemporal relations to a proper subset of themselves — in this case the subset of local simultaneity and conti-nuity along a subset of one-dimensional spacetime paths.

III

At this point it is useful to go a bit further into the dialectic which seems, inevitably I think, to ensue in the course of some such epistemically motivated reductions. The bulk of the debate gener-ally hinges on such questions as the kind and nature of the connec-tion between the accepted primitive elements and the defined con-cepts. Less attention has been paid, unfortunately I think, to the presupposition of the existence of a correct primitive set at all.

As an example of such a dialectic consider the recent debate on the "conventionality" of the metric and of distant simultaneity in relativity. Following the usual presentations of special relativity it is frequently asserted that the metric of Minkowski spacetime and the relations of simultaneity for distant events are matters of conven-tion, for they are not, in this usual presentation defined purely in terms of local simultaneity and causal connectibility.

But reference to the work of Robb, and more modern versions of the same results, shows us that we *can* offer definitions of spatial and temporal separation in terms of causal connectibility alone. Indeed, if we impose some weak constraints on the spatiotemporal relations, they are uniquely so definable. Does this then show that these relations are *not* conventional?[1]

---

1. The Robbian results are contained in A. Robb, *A Theory of Time and Space* (Cambridge: Cambridge University Press, 1914), and *The Absolute Relations of Time and Space,* by the same publisher, 1921. Related and more modernly de-rived results are in E. Zeeman, "Causality Implies the Lorentz Group," *Journal of Mathematical Physics* 5 (1964): 490–493. A clear presentation of the mathemati-cal results, along with an attempted philosophical use of them, is in J. Winnie, "The Causal Theory of Space-Time," in J. Earman, C. Glymour, and J. Stachel, eds., *Foundations of Space-Time Theories, Minnesota Studies in the Philosophy*

Reference is then made to general relativity. Here it turns out that even in the cases where Robbian axiomatic constraints on causal connectibility are satisfied, that is, in spacetimes globally conformal to Minkowski spacetime, there will now be a divergence between spatiotemporal metrics as Robbianly defined and the usual metrics taken for the spacetime. And, in the general case, the Robbian axioms won't even be fulfilled, thus vitiating the very possibility of "causal" definitions of the metric notions in the Robbian vein.[2]

These results show us a number of things. First, the coextensiveness of a "causal" and a spatiotemporal relation will sometimes be a contingent matter. Which relations are coextensive with which others will depend upon just what kind of spacetime we live in. If we take it to be a conceptual and a physical possibility that the spacetime we are in is one in which the coextensiveness required for a particular causal theory of spacetime breaks down, then we will be able to attribute neither physical necessity nor "analyticity" to the propositions asserting that the coextensiveness holds. Then, even if the spacetime in which we live turns out to be one in which the coextensiveness does hold, we will still view this as an inadequate relationship on which to build an epistemically motivated reduction of the spatiotemporal to the causal relation, for we will still view it as lacking the requisite necessity demanded of a reductive definition of this kind.

Exactly the same dialectic can be seen repeated in recent discussions of the possibility of a causal theory of spacetime topology. In Minkowski spacetime it is easy to characterize the manifold topology of the spacetime in terms of a causally defined topology — the Alexandroff topology. In spacetimes of general relativity the same definition will hold, provided that the spacetime is suitably non-pathological in the causal sense. In particular, the identification of the usual manifold topology with the Alexandroff topology will hold just in case the spacetime is strongly causal. If we weaken the causal constraint on spacetimes of strong causality, we can still

---

of Science, vol. 8 (Minneapolis: University of Minnesota Press, 1977), 134–205. See also D. Malament, "Causal Theories of Time and the Conventionality of Simultaneity," *Nous* 11 (1977): 206–274.

2. See L. Sklar, "Facts, Conventions, and Assumptions in the Theory of Spacetime," chap. 3 in this volume.

show that in cases where the spacetime satisfies a somewhat weaker demand of nonpathology, past and future distinguishingness, the manifold topology will still be at least implicitly definable in terms of the notions of causal connectibility.

Where that weaker constraint is violated, however, it is possible to find topologically distinct spacetimes which are alike in their causal connectibility structure, indicating that one cannot, in these cases, ever implicitly define the topology in any terms which take only causal connectibility as primitive. Once again we have the situation that a coextensiveness holds between a spatiotemporal relation and some causal relation in certain familiar spacetimes, but the conceptual possibility of other spacetimes, and perhaps their lawlike possibility as well, the usual laws of general relativity not precluding causally pathological spacetimes, shows that the correlation of spatiotemporal and causal relation is too weak to have the character of analytic necessity implicit in the epistemologically motivated reductionist program.[3]

These failures of the epistemically motivated program of providing causal definitions for spatiotemporal concepts ought not lead us to prematurely abandon the program. Consider, for example, the program to define the topology of spacetime in a causal manner. If we subtly change our notion of what is to count as the causal primitive in the defining base, a far more persuasive version of epistemically motivated causal definition can be constructed.

Instead of taking as one's causal primitive causal connectibility or variants of that notion, take instead the notion of a continuous causal path, in relativity the path of some suitable genidentical causal signal. One can show that in any general relativistic spacetime whatever, the set of continuous causal paths will certainly implicitly define the usual manifold topology. That is, if we have two such relativistic spacetimes, each endowed with a manifold topology, then any one-to-one mapping between them which preserves continuous segments of causal paths will preserve the manifold topology as well.

Even in this case we must be just a little cautious. Other, non-

3. S. Hawking and G. Ellis, *The Large Scale Structure of Space-Time* (Cambridge: Cambridge University Press, 1973), esp. chap. 6. Also D. Malament, "The Class of Continuous Timelike Curves Determines the Topology of Spacetime," *Journal of Mathematical Physics* 8 (1977): 1399–1404, and L. Sklar, "What Might Be Right about the Causal Theory of Time," chap. 9 in this volume.

manifold topologies will still be compatible with the given set of continuous causal path segments, but not be homeomorphic to the usual manifold topology. It would be at this point that a defender of the causal theory we have in mind would likely move, with some plausibility, to the line that the difference between the usual manifold topologies and these new, nonstandard, topologies, unlike that between two distinct manifold topologies, is merely a matter of conventional description of the same "real" topological facts, these taken to be exhausted in our characterization of continuity along causal paths.[4]

IV

Reflection on the dialectic illustrated by the above examples shows us, I believe, a number of important things about epistemically motivated causal theories.

First, there is much we would need to say in justifying any such reduction concerning the status of the propositions which serve to define the spacetime notions in terms of the causal. Merely showing that in some particular spacetime there happens to be a coextensiveness between some spatiotemporal feature and some causal feature won't do to establish the definability of the former in terms of the latter. Not even lawlike coextensiveness will do, I believe. For the reduction to be acceptable to us we want the connection to have that kind of necessity traditionally associated with analyticity. We want to be convinced that, in some reasonable sense, the causal notion really captures "what we meant all along" by the spatiotemporal notion.

Consider for example the critique of the Robbian causal definitions of metric notions in Minkowski spacetime. Even if we believed spacetime was Minkowskian, even if we believed that this was a matter of law, we would, I maintain, be reluctant to accept Robb's "definitions" as definitions of the spacetime metric notions. They would, we think, grossly misrepresent the verification procedures commonly associated with these notions and, for the familiar reasons associating meanings with primary conditions of

---

4. E. Zeeman, "The Topology of Minkowski Space," *Topology* 6 (1967): 161–170, describes the nonmanifold topologies. On the determination of the metric by causal curves and the philosophical relevance of this, see the items by Malament and Sklar in n. 3 above.

warranted assertability, so misrepresent what we meant by the concepts. A definition for distant simultaneity which resorted, as Robb's do, to conditions of causal connectibility arbitrarily far out in the spacetime simply doesn't seem to capture what we meant by distant simultaneity, or rather the closest thing to what we meant by it prior to realizing the existence of the newly discovered constraints on the physical possibilities of verification, all along. But the Einstein definitions do have just that virtue lacking in Robb's. For that reason many would say that they correctly define the metric notions, even if this leads some who accept these definitions to submit to the claim that accepting them imposes the burden of relegating metric features to the realm of conventionality.

Accepting the above means, of course, accepting some sort of analytic-synthetic distinction among the propositions of one's total theory. Just how much of the positivist creed one need swallow here is another story, one I would rather not pursue here.

Next, it would be useful to reflect somewhat on the reduction basis presupposed by the epistemically motivated causal theories we have had in mind. In all of the cases at which we have looked, the same items in the reduction basis come up again and again: simultaneity at a point, causal connectibility and continuity along causal (timelike or lightlike) paths. Since the second can plainly be defined, in the relativistic context if not in general, in terms of the third, events being taken to be causally connectible just in case there is at least one continuous causal path containing them, we may focus on the first and third notion. While there may be some intrinsically causal aspect to the third, the first, local simultaneity, seems itself a spatiotemporal notion with no particular "causal" nature at all. The third also on reflection looks more intrinsically spatiotemporal than causal. If being a continuous causal path means being the path of some genidentical particle, and if genidentity is unpacked itself in terms of spatiotemporal continuity, it seems that it is this notion of a continuous one-dimensional path in the spacetime which is primitive rather than anything specially to do with causality. Of course not all such paths are taken as primitive, only those of a timelike or lightlike nature, and these are, indeed, the paths, in the theory, of causal connection. But is it *that* aspect of them which leads us to place their continuity and discontinuity in the defining basis? Isn't it, rather, that they are the paths "in principle" traversable by the experience of an observer, and, hence, the paths whose continuity or discontinuity could be determined in a "direct," "noninferential," and "theory independent" way? Perhaps the title of causal theory of spacetime is really a misnomer for the epistemically motivated theories we have been examining. On reflection, they seem, rather, to be theories which attempt to reduce the total structure of spacetime to the structures characterizable in terms of a proper subset of our full set of spatiotemporal concepts, the reduction being rationalized by the claim that our full epistemic access being limited to such things as local simultaneity and continuity along causal paths, it is only these relations which we should take as bearing the full load of the "real" nature of spacetime, the rest to be confined to the realm of, at best, the defined and, at worst, the merely conventional.

This brings us to our last, and most important, reflection on the structure of these epistemically motivated reductions. The fundamental problem with all such reductions is that they always seem to go too far. We would like to eschew absolute space with the neo-Newtonians, abandon global inertial frames with those who invoke curved spacetimes instead of flat spacetimes plus gravitational fields, and abandon a nonrelativized notion of simultaneity with relativists. But the epistemic critique which allows us to do so, basically a unity with the kind of epistemically motivated causal theory we have been discussing, seems, when carried to its rational limit, to take us too far.

If we really are to confine ourselves in the reduction basis to the directly apprehended, noninferential, non-theory-loaded realm, and if the privileged spacetime concepts do not have these features and are not in that realm, then they lose their status as the correct reduction basis. We rapidly find ourselves slipping into the attribution of conventionality to many of our formerly most treasured spacetime features. Beginning by eschewing substantivalism for relationism, we find ourselves continuing on to an account of the reality of spacetime which is at best phenomenalistic and at worst solipsistic.[5]

The problem is, of course, a familiar one, and one far more general than merely a problem with theories of spacetime. If we fail to make a distinction in kind between that which is epistemically

5. For some early, and not very satisfying, reflections on this, see H. Reichenbach, *The Axiomatization of the Theory of Relativity* (Berkeley: University of California Press, 1965), Introduction.

accessible independently of hypothesized theory and independently of inference and that which is not, it becomes hard to see how we can carry out the epistemic critiques and attributions of theoretical equivalence we want to maintain. We *want* to say that Heisenberg and Schrödinger are merely two representations of one and the same quantum theory, we want to prefer, on Ockhamistic grounds, neo-Newtonian spacetime to Newton's, curved spacetime to flat plus gravity and relativity to aether theories. Yet all the arguments in favor of these preferences rely, ultimately, on an observational/nonobservational distinction of a more than merely contextual and pragmatic nature.

There are, of course, those who would deny the very intelligibility of the fundamental distinction here. The litany is familiar: all observation is theory laden, direct apprehension is a myth, facts are "soft" all the way down, etc. But from that point of view it is impossible to see the point of a causal theory of spacetime of the kind we have been considering in the first place, and, indeed, hard to see what the point is of the epistemic critiques "built in" to many of the most fundamental physical theories we do accept.

The problem is this: If we are, say, going to eschew an absolute notion of simultaneity for distant events, resting our disavowal on an Einsteinian critique, then we ought to pare our reduction basis down to those spatiotemporal and causal concepts which truly deserve our respect as being purged of conventionality. But then it is hard to see why we ought to stop at local simultaneity and continuity along causal paths if these notions are themselves construed as outside the realm of noninferential direct access. And if it is relations in this realm which are the real reduction basis, then are they not relations in the realm of private experience, in the realm of phenomenal space, and not intersubjectively determinable physical relations at all?

I have no intention whatever of resolving any of these perplexities here, nor even of surveying the familiar collection of solutions or of allegations to the effect that the perplexities are merely the result of typical philosopher's confusion. Here I only want to emphasize the inevitability of these questions once a causal theory of spacetime of the epistemically motivated sort is entertained. What we shall see shortly is that a related, if subtly different, problem arises when we follow out the other variety of causal theories of spacetime to their fatal limits.

V

We saw that the term 'causal' used in characterizing the epistemically motivated causal theories of spacetime was, perhaps, a misnomer. While causal theories might play a role in characterizing the primitive observation basis, the fundamental nature of the relations in that basis was itself prima facie spatiotemporal, although, of course, the relations in the basis are a proper subset of the full array of spatiotemporal relations. The other approach to causal theories is also somewhat misdesignated as 'causal', but for different reasons.

The theories we are concerned with now are those which allege that some or all spatiotemporal features of the world can be considered reduced to some other features, themselves not prima facie spatiotemporal, in the same sense in which material objects reduce to their molecular and atomic constituents, light reduces to electromagnetic radiation, lightning reduces to atmospheric ionic discharge, etc. In other words, the claims will be that there is a reduction of the spatiotemporal to the not prima facie spatiotemporal which is a reduction by means of property identification.

One example of such a claim is, I think, the theory that the direction of time reduces to the (overall) direction of entropic increase. Since it is the direction of entropy increase, rather than the direction of causation, to which the direction of time is to be reduced, there is something misleading about calling this a causal theory. Indeed, in one such version of the theory, Reichenbach's, the direction of causation itself is "reduced" to that of time and ultimately to that of entropic increase.[6]

The nature of this alleged reduction has been sometimes misunderstood, by myself among others. The fact that it is meant to be a reduction by means of scientifically established identification vitiates any objection to the proposed reduction which relies on some epistemic priority for the temporal direction as directly or more

6. The origins of the entropic theory can be found in L. Boltzmann, *Lectures on Gas Theory 1896–1898,* trans. S. Brush (Berkeley: University of California Press, 1964), 446–447. The fullest account of the theory is H. Reichenbach, *The Direction of Time* (Berkeley: University of California Press, 1956). For some critical objections to the account, see J. Earman, "An Attempt to Add a Little Direction to 'The Problem of the Direction of Time'," *Philosophy of Science* 41 (1974): 15–47. Some criticisms of my own, which I now think largely misdirected, are in L. Sklar, *Space, Time, and Spacetime* (Berkeley: University of California Press, 1974), 404–411.

immediately perceived than the entropic. After all, there is a sense in which we know that there is light before we know that there are electromagnetic waves. Optics nonetheless reduces to electromagnetic theory and light waves are identical to a kind of electromagnetic wave.

The entropic theory of time direction is a curious sort of identificatory reduction. It is not some class of things which is being reduced to a class of things, but a relation which is being reduced to a relation. Indeed, most versions of the theory are even a little more peculiar, for it is only a part of a relation which is being reduced. In standard versions of the theory some temporal relations are simply presupposed as primitive. It is only the asymmetry of the relation which is ultimately to be accounted for entropically. Again the theory is subtle for other reasons. It is the *general* direction of time which is established entropically, and this only in a statistical way. Not every pair of events which bear the 'afterward' relation the one to the other need bear any entropic difference the one to the other. Indeed, entropy may be out of line entirely as characterizing either event. In Reichenbach's version of the theory entropy establishes time asymmetry for some pairs of events, and the asymmetry is projected onto other pairs by means which aren't entropic in nature.

The main point is this: Time direction is supposed to reduce to the "causal" notion of direction of entropic increase. But the justification of the reduction is not like that which rationalizes the epistemically motivated reduction. No claim is made to the effect that our sole epistemic access into the asymmetry of time is through awareness, in a more direct sense, of entropic states and their relative magnitudes, nor is there some more direct awareness of the underlying statistical facts about randomness of microstates. Rather it is an empirical discovery that the relation in the world which "underlies" our sense of the direction of time is that of the temporal direction of entropy increase.

Just as one would establish the truth of the claim that light waves are electromagnetic waves by showing that such an identification would account for all the observed and theorized features of light, so this theory is justified by an attempt to show that everything we take to be the case about the asymmetry of time (or, rather the asymmetry of the world in time), can be accounted for in entropic terms. Hence Reichenbach's assiduous attempt to show us that the fact that we have records and memories of the past but not of the future, that we take causation as going from past to future, etc., all have their origin in the prevalent increase of entropy of otherwise unmolested branch systems. Hence also his project, never carried out, of trying to show that entropic increase was also at the root of our subjective "direct" awareness of the asymmetric relation in time of events of private experience.

I am not in the least, of course, maintaining that this theory can be successfully established. Only that if it would be it would be a reduction of a spatiotemporal notion to some other not at all like the earlier causal theories of the epistemically motivated sort. Without wanting to demand perfect symmetry for the analogy, this causal theory of time (or of time asymmetry) is to those discussed earlier much as the theory that tables are molecules is to the theory that tables are logical constructs out of sense-data.[7]

Another example of such an "identificatory" reduction of spacetime relations to relations not prima facie spatiotemporal is even more dramatic in its claims. Unfortunately, it can't be described in any real detail, since, at the moment, it is rather a speculative hope than a genuine reductive program. The idea here is this: we start with a collection of events which are to be the results of measurements or quantum systems. We look for algebraic interrelationships of the results of distinct measurements imposed by the quantum laws. We then hope to "reconstruct" the spacetime structure on events by identifying it with some abstract structural interrelations among measurements algebraically described. The topological, manifold, and metric features of spacetime are dropped as primitives, only to reappear as "identical" with the underlying combinatorial causal relationships.[8]

Notice that in both these examples and indeed in any such attempted reduction by identification, there is no attempt to show that the spatiotemporal relations don't exist. In showing that light is electromagnetic radiation we aren't attempting to show that light does not exist either. It is just that the reduced entity or relation is identified with the reducing. What constitutes the *asymmetry* of

7. More on this is contained in "Up and Down, Left and Right, Past and Future," chap. 12 in this volume.
8. Perhaps the fullest attempt at a reduction of spacetime to algebraically characterized relations among quantum measurements is in D. Finkelstein's set of pieces on the "space-time code." *Physical Review* 184 (1969): 1261; *Physical Review D* 5 (1972): 320; 5 (1972): 2922; 9 (1974): 2219; and D. Finkelstein, G. Frye, and L. Susskind, *Physical Review D* 9 (1974): 2231.

the reduction is an interesting question, given the symmetry of the identity relation. In the case of the identification of things with arrays of atoms it is presumably the asymmetric part-whole relationship. In the case of the identification of light with electromagnetic radiation, it is presumably the greater generality of the reducing theory vis-à-vis the reduced. There are lots of electromagnetic waves which are not light, but all light is electromagnetic waves, for example. What would motivate the asymmetry in the case of the entropic theory of time asymmetry or in the case of the algebraic-combinatorial theory of spatiotemporal relations in general I won't pursue.

VI

At this point we should note some general features of causal theories of spacetime which are meant to be identificatory reductions, and we should contrast those features with some we noted of reductions which have the epistemic motivation.

We saw how, in the cases of epistemically motivated causal theories, there was an interesting mixture of aprioristic and aposterioristic elements. What counted as a spatiotemporal or causal feature to be included in the reduction basis was determined, a priori, by our views regarding features of the world available to our cognition without inference or mediation of theory. But the extent and nature of additional meaningful spatiotemporal concepts, as well as, at least in part, the distinction between the factual and conventional portions of our spacetime picture, depended upon the possibilities available for verification procedures. Since these were usually taken to be possibilities in the mode of physical possibilities, it was then an a posteriori matter just how the reduction basis could be extended out by possibilities of verification into a full spacetime account. That "afterness" for events in each other's light cones is a "hard fact" and that the simultaneity of distant events is "merely conventionally defined," for example, indeed that any such notion of distant simultaneity must needs be inertial frame relative, these are conclusions which follow from the factual matter of the existence of a causal signal of limiting velocity.

There is less apriorism in the kinds of spacetime theories we are now exploring. Proponents of these theories are likely to argue that just as there is nothing in the least bit a priori about the fact that light waves are electromagnetic waves or table arrays of atoms, so it is a purely empirical matter that the direction of time is that of entropic increase or that the structure of spacetime in general is merely the algebraic combinatorial relations among actual and possible quantum measurements.

Nonetheless, it is still possible, even given this view, to demand of the propositions connecting spacetime structures to their reduction bases a *necessary* status. At least that is possible if one makes the familiar distinction between metaphysical necessity and epistemic aprioricity, and also goes along with recent popular claims that genuine identity statements are necessary in the metaphysical sense.

Consider, for example, the claim that the direction of time is that of entropic increase. Presumably, we could not have found this out a priori. It certainly doesn't seem to be "intuitively" true, nor does it seem analytic in the usual sense. Whatever we meant, all along, by the future time direction, it doesn't seem too plausible to claim that we had in mind, all along, the direction of entropic increase — in the sense in which a phenomenalist would argue that in speaking of material objects all along we had in mind, really, merely talk about actual and possible sensory experiences.

Yet, could there be a possible world in which entropy decreased in time (*pace* all the usual qualifications and subtleties here)? Perhaps not. Just as it has been claimed that in any possible world water is H_2O and light is electromagnetic radiation, perhaps a case could be made out that in any possible world the future direction of time just *is* the direction of entropic increase. And in any possible world, one might claim, the spacetime structure just *is* the same combinatorial structure on measurements it is in the actual world. Of course, the argument continues, just as there could be worlds where something other than water had all the features which we took in this world as the "stereotype" or "reference fixing features" of water in this world, so, presumably, there could be worlds where what presented itself to us in the way spacetime does in this world was something different. But there does seem something spurious about this. It is a matter I will come back to.[9]

9. On the necessity of identification statements, see S. Kripke, "Naming and Necessity," in D. Davidson and G. Harman, eds., *Semantics of Natural Language* (Dordrecht: Reidel, 1972), 253–355; and H. Putnam, "The Meaning of Meaning," in K. Gunderson, ed., *Language, Mind, and Knowledge, Minnesota Studies in the Philosophy of Science,* vol. 7 (Minneapolis: University of Minnesota Press, 1975), 131–193.

VII

Now one could argue against any particular identificationist causal theory of spacetime in many ways. But the argument which is most important for us is one which, if it is correct, allows us to dismiss in advance the possibility of the correctness of any such identificatory causal theory of spacetime. In the particular case of the direction of time it has been expressed with some forcefulness, if less clarity, by Eddington. We know, he says, what the nature of the afterward relation is like, directly and noninferentially, from our immediate experience. We also know what the relationship is like which is that of one state of a system having greater entropy than another. And we know that these two relations are "utterly different." Neither "afterness" (Eddington speaks of "becoming") nor having a more randomized microstate can be viewed as "somethings I know not what" in a kind of Ramsey sentence way of looking at theories, a mode of being picked out sufficiently unspecific to allow us to find out that the state in question does indeed have some nature utterly unexpected. In Eddington's language, "We have direct insight into 'becoming' which sweeps aside all symbolic knowledge as on an inferior plane."[10]

The identificationist's likely reply to this move is also clear. We have a direct, nonsymbolic knowledge of hotness and coldness also. If Eddington's arguments were correct, ought we not to say that hotness and coldness could not be states of higher or lower mean kinetic energy of micro-constituents? Yet that is exactly what those temperature states are. Of course, such a line would usually continue, there is "felt hotness" and "felt coldness," but these are mere "subjective states of awareness of the observer." Hotness and coldness in the object are more properly thought of as "symbolically" known through their effects, and it is these objective features of nature which are identified with the micro-states in the usual reduction of thermodynamics to kinetic theory.

But now, in the light of all this, just how plausible do our identificatory reductions of spatiotemporal features appear? Presumably, here too the refutation of the Eddingtonian will require that "afterness" as immediately apprehended, or, in the more general case, all spatiotemporal features as we know them in our "mani-

10. A. Eddington, *The Nature of the Physical World* (Cambridge: Cambridge University Press, 1928), chap. 5, "Becoming."

fest" world of direct experience, are, as in the case above, to be "sliced off" from the physical world and disposed of in that convenient receptable "the subjective awareness of the observer." The afterness of the world, or its spatiotemporality is then as distinct from felt afterness or sensed spatiotemporality as agitated molecules are from the *felt* sensations of hot and cold.

That we are pressed in this direction by identificatory causal theories of spacetime which attempt to say that spacetime is nothing but some non-prima-facie spatiotemporal structure in the world has an interesting historical antecedent. Meyerson argued that the direction of science was always to eliminate diversity in terms of unity. One such program would be that which attempted to eliminate the sensed diversity of the world in terms of a reduction of all other features to the spatiotemporal. This would be the Cartesian program in any of its older or newer guises (of which geometrodynamics is perhaps the most well known current version). All there is are changing modes of extension. He pointed out that in such a reduction of all that is physical to modes of extension, it was inevitable that much of the directly sensed diversity of the world woul be expelled from consideration in the reductionist program by being conveniently ejected out of the physical world and into the "merely subjective" realm of private experience. Here we see that exactly the same pressures work when spacetime is to be identified as some structure not prima facie spatiotemporal, as hold when the not prima facie spatiotemporal is identified with modes of extension.[11]

But the pressures may be even more disturbing here. It is one thing to dump sensed color, felt heat, and all the other "secondary" qualities into the realm of the mental, reserving as objective only extension and its modes and all those other, "symbolically known," features which can be identified with the modes of extension. It is, at least to some, even more disturbing to have to count spatiotemporality itself, at least as it is "directly present" to us in our manifest world of awareness, as merely secondary, leaving in the actual world spatiotemporality only as a "symbolically known"

11. E. Meyerson, *Identity and Reality,* trans. K. Loewenberg (London: George Allen and Unwin, 1930), esp. chaps. 11 and 12.

set of relations identical in nature with some relations among states of affairs not prima facie spatiotemporal at all.[12]

VIII

What are the prospects of a causal theory of spacetime? As we have seen, there are two quite different schemes one might have in mind when advocating a causal theory. One might be proposing a reduction of spatiotemporal relations to a proper subset of themselves (and perhaps some genuinely causal relations as well), motivating the reduction by an epistemic critique which, fundamentally, identifies meaning with mode of verification. Here there will be problems regarding the association of reduced with reducing motion, the familiar problems of justifying a claim to analyticity for some proposition. More crucially, the reductionist theory will seem to drive one to a reduction basis not in the objective world at all, leaving one puzzled as to how one's causal theory of spacetime could really be a theory of the physical world at all.

On the other hand, one might instead be claiming that, as an empirical matter of fact, the spatiotemporal features of the world are identical with other features not prima facie spatiotemporal. Here the fundamental difficulty will be that in carrying out the identification in a plausible way we will be required to shunt into the realm of "mere subjectivity" the spatiotemporal features of the world as we encounter them in our immediate experience. Once

12. The fullest attempt to overcome the difficulties incurred by the scientific-identificatory reductions of time to causality of which I am aware is that of H. Mehlberg in his "Essay on the Causal Theory of Time." Originally published in 1935–1937 in French, the work is now available in English translation in the posthumously published *Time, Causality, and the Quantum Theory* (Dordrecht: Reidel, 1980). Mehlberg accepts the disassociation of experienced from physical time which is, it would seem, an unavoidable consequence of such a scientific-identificatory causal account. He attempts to mitigate this consequence, unsuccessfully I believe, by invoking a "universal" time which subsumes the subjective and physical times. The frequent congruence of these times is alleged to be a consequence of psycho-physical parallelism of the realms of immediate experience and the physical realm of the brain. The doctrine, though, requires one to make sense of the notion of simultaneity of physical and mental events. Since this simultaneity can, it would seem, be a simultaneity neither in the sense of subjective nor in the sense of physical time, it isn't clear to me just how the doctrine can be coherently formulated. In the "Essay" see esp. Part II, and most esp. chap. 10 and the Supplement.

again difficult questions regarding the relationship of what we immediately apprehend to what is objectively present loom.

There have been, of course, numerous attempts to relieve us of the two anxieties into which we have been led. Perhaps some are convinced of their success. For my own part I believe that I have yet to see an account which will: (1) do justice to the persuasiveness of the epistemic critique as it appears in functioning science, particularly in the critical evolution of spacetime theories in recent years; (2) do justice to the persuasiveness of identificatory reduction as a means of achieving unification of theory, including its persuasiveness in such contexts as Boltzmannian — Reichenbachian identifications of the future direction of time with the direction in time of entropic increase; and yet (3) which will neither trap us in a solipsistic prison nor remove from the world of objective reality such essential features as existing in space and time *as we know them.* The prospects for the success of a causal theory of spacetime are no better or worse than the prospects in general of an adequate account of the relation of the "scientific" to the "manifest" world.

11. Time, Reality, and Relativity

There are venerable doctrines to the effect that time is radically unlike space. Some of these doctrines rest upon alleged asymmetries in the way in which temporality and spatiality are connected to existence. While being elsewhere hardly implies not really being at all, some have alleged that only present entities are real, past and future not having genuine existence at all, and others have maintained that only the present and past are real, denying full reality to the yet to be.

Others have maintained that the acceptance of special relativity, with its integration of space and time into Minkowski spacetime, plainly refutes any such claims concerning the unreality of the past and future or of the future.

This chapter explores the legitimacy of the contention that one can't have both special relativity and the irreality of the nonpresent or of the future. First, some motivations behind the doctrines of irreality are examined. Next, several alternatives open to one who accepts special relativity are outlined. While it is agreed that accepting relativity places constraints upon one's doctrine of irreality, essentially making one treat the elsewhere more symmetrically with the elsewhen than was congenial to the older irrealist doctrines, I maintain that one cannot simply "read off" simple metaphysical conclusions from the relativistic theory. Rather, the metaphysical stance one ought to adopt follows only from the adoption of a number of fundamentally philosophical postulates. Further, a careful examination of special relativity itself will show that how we ought to view that theory itself is subject to philosophical critique.

I

THERE is a doctrine, venerable and very familiar, that that which does not exist in the present does not, properly speaking, exist at all. Alternatively there is the equally ancient and equally intuitive view that only the past and present have determinate reality and that the future has no such being, or at least no such determinate being.

For the moment I don't wish to explore the fundamental questions to which these doctrines give rise: "Why do they have the intuitive appeal they do?" "Can any good reasons whatever be given to support them?" "In the final analysis could they possibly be correct?" Rather I want to take a look at what purports to be a simple and conclusive refutation of all such doctrines. For, it has been claimed, "science" refutes the asymmetric treatment of the present and nonpresent once and for all. While this argument too has been "floating around" for some time, it has fairly recently appeared in the literature, once in a version replete with infelicities of expression and formulation (talking about determinism of the future when it is the question of determinate reality which is at issue) and the other time framed with greater philosophical sophistication. But in both cases the argument is fundamentally the same.[1]

Consider an observer at a place-time. According to the doctrine in question, events in his future (say) are not determinately real. But according to relativity there is going to be another observer, coincident with the first, and hence certainly real to him, since immediately present to him. Now many of the events future to the first observer will be present to the second, so long as the two observers are in relative motion. Indeed, for any future event (relative to the first observer and spacelike separated from him) there will be a second observer such that that event is present to the second observer when the second observer is coincident with the first. So the "future" event will be real, relative to the second observer. But surely "being real to" is a transitive notion. If the event is real to the second observer who is real to the first, it must be real to the first observer, contradicting our original claim that events future to an observer lack reality for him.

Once we have accepted the principle of transitivity of reality then we can go further. For even events in my future light cone will be present, hence real, for some observer who is present, hence real, for an observer in motion with respect to me but coincident to me

1. C. Rietdijk, "A Rigorous Proof of Determinism Derived from the Spatial Theory of Relativity," *Philosophy of Science* 33 (1966): 341–344. H. Putnam, "Time and Physical Geometry," *Journal of Philosophy* 64 (1967): 240–247.

and, hence, real for me. So even events in my absolute future must be declared to have determinate reality.

Now obviously the whole argument rests upon the fundamental assumption of the transitivity of 'reality to'. Given the relativization of simultaneity to a reference frame in relativity, anyone who wishes to relate determinate reality to temporal presence must also relativize having reality to a state of motion of an observer. And given the nontransitivity of simultaneity in relativity across observers in differing reference frames, we could also easily find our way out of this argument by simply denying that 'having reality for' is a transitive relation. Now Putnam calls the transitivity of 'reality for' the principle of there being "No Privileged Observers," and, surely, we would like all observers to have equal rights to a legitimate world-description.[2] But why one would think that such a doctrine of "No Privileged Observers" would lead one immediately to affirm the transitivity of 'reality for', given that one has already relativized such previously nonrelative notions as that of simultaneity, is beyond me.

But simply blocking the argument against the traditional doctrine in this way is of very little interest. For example, we still are at a loss as to what specific doctrine about reality we should adopt in the relativistic case. Should we simply relativize the old doctrine taking, as before, the present to be the real and simply denying the transitivity of 'reality for'? Or should we adopt some alternative, more radical, view? To decide requires that we look a little more closely into the metaphysical presuppositions which underlie the relativistic spacetime picture itself. What I want to do here, rather than pay attention only to the problem as so far narrowly construed, is to explore the more general issue suggested by Putnam in the concluding remarks to his paper:

> I conclude that the problem of reality and determinateness of future events is now solved. Moreover, it is solved by physics and not by philosophy. . . . Indeed, I do not believe that there are any longer any *philosophical* problems about Time; there is only the physical problem of determining the exact physical geometry of the four-dimensional continuum that we inhabit.[3]

I think that such a naive view is as wrong as can be. Just as a computer is only as good as its programmer ("Garbage in, garbage

2. Putnam, "Time and Physical Geometry," 241.
3. Ibid., 247.

out"), one can extract only so much metaphysics from a physical theory as one puts in. While our total world view must, of course, be consistent with our best available scientific theories, it is a great mistake to read off a metaphysics superficially from the theory's overt appearance, and an even graver mistake to neglect the fact that metaphysical presuppositions have gone into the formulation of the theory, as it is usually framed, in the first place.

II

The original Einstein papers on special relativity are founded, as is well known, on a verificationist critique of earlier theories. Referring to the observational facts, generalizable as the null-results of the generalized round-trip experiments, he argues for the necessity of finding an "operational" meaning to apply to simultaneity for events at a distance and for the impossibility of doing this in any way which allows us an empirical determination of the one-way velocity of light. From then on the moves are all well known which invoke the "radar" method for establishing simultaneity, this resting upon the conventional stipulation of the uniformity of the velocity of light in all directions. Nothing in the way of newly predicted phenomena, such as those predicted by the mechanics designed to preserve the old conservation rules relative to the newly adopted spacetime picture, changes the fundamental point — that quantities are to be introduced into one's theory only insofar as they are empirically determinable or conventionally definable from empirically determinable quantities.

Of course, even within the relativistic context it is easy to forget the verificationist arguments which initiated the theory in the first place, and to forget the distinction within the foundations of the theory between the genuinely factual elements and the merely conventional choices which go into the relativistic spacetime picture. Hence the necessity for frequently reintroducing the claim of conventionality and the difficulty of seeing it through to its full consequences even for those who espouse it. Thus we have the Reichenbach-Grünbaum argument to the effect that in picking ϵ equal to 1/2 in the familiar "radar" definition of simultaneity, a merely conventional element has been introduced, corresponding to picking the velocity of outgoing and returning light as equal. But it then takes some effort to see that even the choice of a linear relationship on the time of emission and reception of a light signal to define the point simultaneous with the light's being reflected at a distant point is itself undetermined by the facts. Thus the very choice of a flat spacetime for the spacetime appropriate to special relativity is easily argued to be as much a "mere conventional choice" as is that of the equal velocities of light in opposite directions.

Now it might be argued that Einstein's verificationism was a misfortune, to be encountered not with a rejection of special relativity, but with an acceptance of the theory now to be understood on better epistemological grounds. There is precedent for an attitude of this kind. Einstein was led to general relativity both by an attempt to satisfy Mach's requirements for an explanation of inertial forces and by the belief that covariance of equations represented a generalization of the relativity principle underlying special relativity. Most would now take the theory itself to be our best available current theory of gravitation but would deny its conformity with Machian requirements and would deny the legitimacy of identifying covariance with a relativity principle of any kind. Even the principle of equivalence, another "background principle" to general relativity, is questionable in a way the theory is not.

But I don't think a position of this kind will work in the present case. I can see no way of rejecting the older aether-compensatory theories, originally invoked to explain the Michelson-Morley results, without invoking a verificationist critique of some kind or other. And I know of no way to defend the move to a relativized notion of simultaneity, so essential for special relativity, without first offering a critique, in the same vein as Einstein's, of the pre-relativistic absolutist notion, and then continuing to observe that even the relativistic replacement for this older notion is itself, insofar as it outruns the "hard data" of experiment, infested with a high degree of conventionality.

III

Once adopt a verificationist stance of any kind and certain fundamental questions arise. First of all there is the question of just what properties and relations are going to be taken as epistemically available by "direct inspection."

Obviously one such relation is that of coincidence between events. Without taking this as epistemically available to us, the whole project of providing an "operational definition" for simultaneity for distant events is blocked at the start. For we must be able to synchronize clocks at a point (determine the emission and reception time of the light beam, etc.), and this amounts to a determination of spatiotemporal coincidence. It is less frequently noted that another fundamental notion is taken as primitive in the definition as well. This is the notion of continuity along a causal (timelike or lightlike) path. The definition requires the use of the times of emission and reception of a reflected light beam. But how could these be determined unless there were available to us some method of determining that the light beam whose later reception time is determined is the very same light beam whose emission time was earlier recorded and whose reflection at the distant point coincided with the event at a distance which is to be identified as simultaneous with some local event? And if, as seems the only plausible move to make in this case, we take identity of the beam through time (or genidentity of the set of events making up the beam) to be constituted by the spatiotemporal continuity of the beam (and what else would serve in its place?), this presupposes that a determination of such continuity is epistemically available to us.

Now there are many absolutely crucial questions to be asked here. Can any coherent sense be made of the claim that our spatiotemporal knowledge is exhausted by reference to data formalizable in the two notions of coincidence and causal continuity? Can an intersubjective physical theory be formulated in terms of such "immediately accessible" concepts at all? To what extent is the restriction of epistemic accessibility to such a limited class of notions in any sense a commitment to a denial of "reality" to those features of the world whose description outruns the capacity of the meager basis to which we have limited ourselves? We will return to some of these questions shortly when we ask ourselves what the consequences are for a doctrine of the "irreality of the past and future" of such an epistemic foundation for relativity.

For the moment, however, I want to digress a little to show how in a slightly different, but closely related, context the natural choice of these two concepts as primitive once more suggests itself.

Even more primitive as a feature of our spacetime than metric features, like simultaneity, is its topology. Both the Minkowski spacetime of special relativity and the pseudo-Riemannian spacetime of general relativity have determinate topologies — the usual manifold topologies. To what extent are these topologies empirically determinable and to what extent, once again, are they merely conventionally stipulated? A natural suggestion has been to look for the source of our epistemic access to topology in the causal structure among the events in the spacetime.

What we soon discover is this: In Minkowski spacetime there is, indeed, a topology definable solely in terms of the causal connectivity among events, the Alexandrov topology, which is provably equivalent to the ordinary manifold topology. Furthermore, the Alexandrov topology is "natural" enough that one might be inclined to the view that what we really meant all along by the topological structure of the spacetime was that which the Alexandrov topology gives us. Essentially, it identifies the basis of open sets of the topology with sets of events timelike accessible from a pair of events, i.e., an open set in the basis is the common region of the interior of a forward light cone from one event and the interior of the backward light cone from another.

When we go to the general relativistic picture, it is plain that there are pathological spacetimes which are such that the Alexandrov topology and the manifold topology will not even be extensionally equivalent. Only if the spacetimes are what is called strongly causal will the two topologies coincide. Indeed, it is easy to show that no topology defined in terms of causal connectibility in any way will be generally adequate even to capture extensionally the usual manifold topology of the spacetime.

A recent result of Malament's does show us, however, that if one assumes the spacetime to have some manifold topology or other, then the topology along the causal paths (timelike and lightlike paths) does completely determine the topology of the spacetime, in the sense that any one-to-one mapping from one spacetime to another which preserves continuity along causal paths will be, relative to the assumption that both spacetimes have the usual manifold topology, a homeomorphism.[4]

But why assume the spacetimes have a manifold topology of the usual sort? Are there any other topologies which one can imagine which will differ from the manifold topologies but which will agree with them regarding the structure of continuity along causal paths? There are. For example, the topology which takes the maximum number of sets as open sets (the finest topology on the spacetime) compatible with the continuity structure along causal paths will agree with the usual manifold topology along these paths and yet be nonhomeomorphic to the manifold topology.

Now why are any of these mathematical results of any philo-

4. D. Malament, "The Class of Continuous Timelike Curves Determines the Topology of Spacetime," *Journal of Mathematical Physics* 18 (1977): 1399–1404.

sophical interest? Clearly, the usual epistemic presuppositions are being made: in order to attribute a structure to the world we must indicate how that structure can be empirically determined by us. The full topological structure of spacetime is not the sort of thing open to any kind of direct or immediate epistemic access. But the continuity structure along causal (lightlike or timelike) world-lines is. Hence the topological structure of the world is determined only to the extent that it is fixed by the continuity structure of the causal world-lines. And why would one think continuity along causal world-lines open to us? Only, I think, because of the implicit claim that they can be "lived along" by observers who, in their traversal of the world-lines, could directly determine the continuity or discontinuity of a set of points.[5]

IV

Let us now return to our original question: to what extent is the old doctrine of the unreality of the past and future undermined by the adoption of a relativistic spacetime picture of the world? Alternatively, can a metaphysics be constructed which retains the old "intuition" and which is compatible with the new spacetime view? And if the latter is the truth, what modifications in the old view must we make in order to retain this compatibility? And how plausible, given the original standpoint of the "irrealist" metaphysics, are these modifications going to be?

Now one obvious way to reconcile irrealism with relativity is simply to drop the principle of the transitivity of 'reality for', and retain the original doctrine in the form most similar to its original version, that is, taking as real for an observer all and only those events which are temporally present to him. Of course we must now relativize 'reality for' so that it is just as nontransitive across observers in mutual motion with respect to one another as is simultaneity, but there doesn't seem to be anything very objectionable a priori about this. Making this move certainly doesn't seem to be positing any observer as "privileged," as Putnam alleges when he calls the principle of transitivity of reality the principle that There are No Privileged Observers. For, just as with simultaneity in special relativity, all inertial observers are on a par and none are singled out as in any way "privileged."

5. L. Sklar, "What Might Be Right about the Causal Theory of Time," chap. 9 in this volume.

But there does seem to be *something* wrong with this approach. The source of our skepticism about it lies, I think, in the strong pressures toward a conventionalist, and, hence, in a sense, irrealist, theory with regard to simultaneity for distant events. If the totality of our epistemic access to the theory is contained in the facts about coincidence and continuity along causal paths, can we reasonably take a realistic attitude toward relations not totally reducible in terms of these notions at all? Of course we can *call* two spacelike separated events simultaneous or not, picking the relativistic definition for distant simultaneity or some other. We can speak of light as having an isotropically distributed velocity, or not, again choosing either the standard relativistic convention or some other. But if any one of these accounts, framed in superficially inconsistent terms, can explain equally well all the hard data of experience, why should we take the accounts as differing at all in the real features they attribute to the world? We are easily led to the (standard) conventionalist claim that there is no fact of the matter at all about the equality of the velocity of light in all directions, and no fact of the matter at all about which distant events are "really" simultaneous with a given event.

If we now associate the real (for an observer) with the simultaneous for him, we must, accepting the conventionality of simultaneity, accept as well a conventionalist theory of 'reality for'. It is then merely a matter of arbitrary stipulation that one distant event rather than another is taken as real for an observer. Now there is nothing inconsistent or otherwise formally objectionable about such a relativized notion of 'reality for', but it does seem to take the metaphysical heart out of the old claim that the present had genuine reality and the past and future lacked it. For what counts as the present is only a matter of arbitrary choice, and so then is what is taken as real. At this point one can easily see why one would adopt, instead, the line standardly taken by proponents of relativistic spacetime and declare all events, past, present and future, equally real. For the distinction among them, being reduced to a mere conventional way of speaking, seems far too fragile to bear any real metaphysical weight.

V

But there is an alternative. It is a radical one and one we will hesitate to take when we see the position into which we are being forced. It is, however, not only consistent with the relativistic spacetime picture, but an option well in keeping, from at least one point of view, with at least some of the original motivations underlying the irrealist view about past and future.

Given an observer at a time, what, from the relativistic point of view, should we cast into the domain of unreality? Certainly all the contents of his absolute past and future, that is everything contained in either his forward or backward light cones.

But what about the contents of the world outside of both light cones altogether? The first alternative we looked at discriminated among these events, taking those simultaneous (in the relativistic sense) with the observer at the instant to be real and discarding the remainder as unreal. But such a discrimination seems all too arbitrary. The obvious alternatives are to count everything spacelike separated from the observer as real or to count it all as unreal for him. And surely of these two alternatives it is the latter which best suits the initial motivation behind the irrealist viewpoint as originally construed.

Not that the former option has no plausibility. After all, we do sometimes speak of the region outside the light cones as the region of the "absolutely simultaneous," and if reality is to be identified with temporal presentness this does suggest that all which is now "absolutely present" should be counted in the domain of reality. I don't think there is any way in which one could "refute" this option. We are, after all, revising the older theory of the reality of only the present to fit the new world picture and, without some further constraints on our choices, I suppose the options are up to us.

Nonetheless, this view certainly seems peculiar. Having dismissed as unreal things whose only deficiency is the fact that causal signals from them have taken time to arrive at us now, or that causal signals from us will take some time to arrive at them, it seems very suspicious indeed to promote into the domain of the fully real those things causally inaccessible to us (now) altogether.

But just what were some of the motivations behind the original intuition of the irreality of all but the present? Perhaps if we get some grip on them we will have some guidance in our attempt to rework the theory to fit these changed circumstances.

(1) One source of the old intuition is plainly the fact that our

natural language is tensed. We speak of things now as existing but of the past only as having existed and of the future only as going to exist. But, of course, we can't look here for an explanation of the intuition of unreality, except possibly in a weak psychological sense. For the claim of the irrealist is that this natural way of speaking is not a mere trivial artifact of ordinary language but reflective of some deep metaphysical distinction between the else-when and the now. What we would like is some interesting distinction between past and future and the present which simultaneously explains the felt metaphysical distinction and its representation by means of tense distinctions in ordinary language.

(2) There is the resort to irrealism which holds that an irrealist view of the future is necessary to avoid fatalism. This connects the alleged irreality of the future with an alleged present absence of truth-values for future tensed statements. The claim is made that this absence of truth-values for future tensed statements allows us to avoid the disturbing conclusion that what will be is already the case, and, in the sense that statements about it already have a truth-value, is somehow fated. It allows us to avoid thinking that the future is as beyond our capabilities of changing as are facts about the past.

From this standpoint we obviously want events in the forward light cone to be unreal. From this perspective we will, contrary to the general time-irrealist point of view, want events in the backward light cone to be real, as sentences about them presumably have, now, determinate truth-value. But what are we to do with events outside either light cone?

By our previous argument from the arbitrariness of distant simultaneity, we will probably not want to discriminate among the events at spacelike separation some we take to be real and some unreal. So our only option would be to take all of them as either real or all as not real. Given that irreality, from this motivational standpoint, was invoked in order to prevent there already being a fact of the matter which, now, determined the truth or falsehood of future tensed assertions, there seems to be little to constrain our choice of reality or unreality for those events which are now absent from both our causal past and our causal future.

Suppose we take all events at spacelike separation to be unreal. We will, of course, have to suffer some consequences which, from the "ordinary language" point of view are rather peculiar. For example, there will be events which are now such that they will be in my real past at some future time, but which will never have a present reality to me at all.[6] But we expect that a move to a relativistic picture will force some violence on our ways of speaking and this is no refutation of adopting this way of thinking about things. Nor, from this point of view (the sole aim of which is to prevent the overabundant reality of our to-be-experienced future), do I see any reason why we couldn't adopt the other alternative of simply taking all spacelike separated events as now real. Of course both of these options will have the virtue (if it is one) that two coincident observers, no matter what their relative motion, will, at an instant, attribute reality and irreality to the same regions of spacetime.

(3) The irrealism which we are primarily interested in is not, however, that which attributes reality to past and present and denies it only to the future, but rather that which takes the real to be only that which is present. What is the motivation behind this view, over and above the, possibly merely artifactual, special role of tensed discourse in our natural language?

At least one motivation for the view is to be found in the "epistemic remoteness" of past and future. This ties in the familiar intuition about the irreality of past and future with such familiar verificationist themes as the claim that all propositions about past and future are, unlike those about the present (or, at least, about present immediate experience) "inferential" in nature. And it ties it in with the further move, so familiar with radical verificationist programs, either to reduce statements about past and future to statements about present evidence or, alternatively, to adopt some kind of "criterion" theory of the meaning of statements about past and future, taking their meaning to be fixed by their relation, in terms of warranted assertability, to statements about present experience which are exhaustive of the body of evidential statements for them. From the latter point of view it is fairly clear what the assertion of irreality to past and future amounts to (denial of bivalence to past and future tensed statements, etc.), and from the former it is at least clear why a radical asymmetry of some sort is being maintained between, on the one hand, present tensed statements and, on the other, those tensed past or future.

I certainly don't intend to examine either of these familiar verifi-

6. Putnam, "Time and Physical Geometry," 246.

cationist claims, either in general or in their particular application to statements about past and future. Rather, I want only to explore what the impact of relativistic spacetime thinking ought to be on the programs. Let me also say here that to the objection that it is not these verificationist themes which really underlie our intuitions about the asymmetry of reality in question, I have no reply. I think the objection may be correct, but I am at a loss to imagine what other source to the asymmetry there might be (over and above, of course, the mere fact of asymmetric grammatical forms in our particular natural language).

Now, looking at the asymmetry as presently motivated, the first thing to say, once again, is that there is nothing in the relativistic picture to force the asymmetrist to give up his position. But how shall he frame it so as to fit the new spacetime picture? Once again he could simply relativize everything to the state of motion of an observer, using the relativized simultaneity notion to demarcate present from past and future. But, as usual, this is subject to the objection of arbitrariness and, in addition, from the point of view of our present, epistemic, motivation for the asymmetrical standpoint we have a far more plausible option to take.

If the past and future are to be declared unreal due to their "epistemic distance" from us, what attitude are we to take toward events at spacelike separation from us? The answer is clear. For events at spacelike separation from us are now (although they may not be in the future) totally immune from epistemic contact by us. That is the very fact which in special relativity leads us to the doctrine of conventionality of simultaneity in the first place, and in the context of general relativity leads to the notions of event and particle horizons, deterministic but unpredictable spacetimes, etc.

So surely in this case the option is fixed for us. If we are to take the past and future as unreal due to their epistemic distance from us, surely we are to declare everything outside the light cone as unreal as well.[7]

What is the most interesting here is this: even from the prerela-

7. There is also the position that the past, being forever epistemically inaccessible to us, is unreal, but that the present and future are both real as both are, now, open to present or future inspection. From the point of view I have been proposing I think it evident what the relativistic revision of this doctrine ought to be. Surely the regions outside the light cone are to be lumped with the past into the domain of irreality, leaving only the future light cone and the present event-point in the realm of reality.

tivistic point of view, it isn't the least bit clear why we should have ever treated the elsewhere differently than the elsewhen. At least it isn't clear until we have some backing for the intuition of unreality of past and future which doesn't rest upon the mere fact of epistemic distance. For even from the prerelativistic point of view, events spatially separated from us seem to be as epistemically distant as those in the past or future. Why, even in the prerelativistic situation, would anyone have wanted to deny reality to the past and yet affirm it of the spatially distant? From the relativistic point of view it seems clear that we are forced (not, of course, by logical consistency but only by plausibility of conceptual structure), if we are going to take an irrealist line as motivated by facts of epistemic distance at all, to certainly deny reality as fervently to the spatially distant as we do to past and future. What relativity does, with the invocation of spacetime to replace space through time, is not to force us in any way (Putnam, Rietdijk, and others to the contrary) to reject our irrealist position, but only to symmetrize it in a natural way to include spatial separation on a par with temporal separation.

VI

From the point of view we have been adopting, the doctrine of the irreality of past and future, taken as having a motivation in their epistemic distance, now seems clearer. We first reduce "reality" to the lived experience of the observer; that is, we first fall, following a well-known verificationist path, into solipsism. Then, seeing that our own future experiences and past experiences are as remote *now* from us as the spatially distant, the not immediately sensed, etc., we fall from solipsism into solipsism of the present moment. Reality has now been reduced to a point!

Now obviously we don't want to go along this path. Where to block it is an interesting question. One can challenge the basic tenets of verificationism either with regard to its theory of knowledge or with regard to its semantic theory. One could certainly try to break the connection between the facts of epistemic distance, even if it is acknowledged to exist, and "irreality."

But what role does relativity theory play in all of this? One thing is certain. Acceptance of relativity cannot force one into the acceptance or rejection of any of the traditional metaphysical views about the reality of past and future. It can lead one to see more clearly than one did previously that by parity of reasoning one ought to treat spatial separation on a par with temporal. By forcing one in addition to say some things which seem peculiar in ordinary language it might lead one to move toward one position or another on grounds one didn't have before. But one who wishes to stick by an irrealist position, and is willing to pay the price, is certainly able to do so, all the while accepting the scientific reasonableness of Minkowski spacetime.

Much more interesting is the relationship between relativity and verificationism in general. Certainly the original arguments in favor of the relativistic viewpoint are rife with verificationist presuppositions about meaning, etc. And despite Einstein's later disavowal of the verificationist point of view, no one to my knowledge has provided an adequate account of the foundations of relativity which isn't verificationist in essence.

That one would want to do so seems fairly clear. Let me illustrate with just one problem. On the basis of verificationist principles one takes the attribution of distant simultaneity, of the isotropy of the velocity of light, etc., to be mere conventional decisions. On similar grounds, as we have seen, one can plausibly argue that the very adoption of the standard manifold topology for spacetime is merely conventional, any other topology saving the continuity structure along causal paths serving to "save the phenomena" equally well.

But now consider the problem of so-called indistinguishable spacetimes. These are spacetimes which are counted as distinct by general relativity (they are by no means isometric to one another), yet which are such that no single observer can tell, even given a complete infinite lifetime, which of the spacetimes he inhabits. This could only be determined by a "super-observer" who had access to the collected information of all the observers of the ordinary sort. Is it merely a matter of convention which of a number of indistinguishable spacetimes an observer inhabits?[8]

Clearly, if one takes saving the phenomena as the sole task for which a theory is to be responsible, and a choice among theories

8. C. Glymour, "Topology, Cosmology, and Convention," *Synthese* 24 (1972): 195–218. C. Glymour, "Indistinguishable Space-Times and the Fundamental Group," and D. Malament, "Observationally Indistinguishable Space-Times," both in J. Earman, C. Glymour, and J. Stachel, eds., *Foundations of Space-Time Theories, Minnesota Studies in the Philosophy of Science,* vol. 8 (Minneapolis: University of Minnesota Press, 1977).

which save the same phenomena to be a matter of mere convention, the answer depends on what one takes the phenomena to be
saved to be. The pressure which drives us, verificationistically, to
take the phenomena to be everything which is epistemically accessible to observers in general can also drive us to take them to be
everything which is accessible only to one observer, oneself. This is,
naturally, just the familiar slide from a phenomenalistic to a solipsistic position. Worse yet, should we not, given the epistemic inac-
cessibility *now* of past and future, take the phenomena to be rather
the data of the one observer at the present moment, leading to a
whole new range of indistinguishable spacetimes? Whereas before
we counted as indistinguishable those spacetimes which appeared
the same to any world-line in them, we now take them to be those
which appear the same to single world-points. Surely verificationism has gone too far at this point.

But what I don't know is either how to formulate a coherent
underpinning for relativity which isn't verificationist to begin with,
or how, once begun, to find a natural stopping point for verificationist claims of underdetermination and conventionality.

12. Up and Down, Left and Right, Past and Future

The claim that the direction of time itself is explicable in terms of the direction in time in which the entropy of physical systems increases has seemed exciting and plausible to some and grotesquely absurd to others. Some of the disagreement, I believe, arises from general misunderstanding of just what kind of relationship between temporal asymmetry and entropic asymmetry is intended by the proponents of the reductive standpoint.

Two comparison cases are examined: the association of left-right asymmetrical systems with asymmetries in weak-interactions, and the association of the spatial directions of up and down with the local gradient of the gravitational field. While the first would not lead a rational person to suggest that left and right reduce in any sense to features of weak interactions, the reasonable person will conclude that the familiar up-down distinction does reduce to features of the world characterizable solely in terms of the local direction of increase of the gravitational field. To which case should we appeal when trying to understand the relation of temporal asymmetry to entropic asymmetry?

The claim made here is that the alleged reduction of temporal to entropic asymmetry is supposed to be a reduction of a kind similar to that accomplished by identificatory reductions in science, not the kind of reduction, epistemically motivated, familiar to philosophers from such examples as the alleged phenomenalistic reduction of objects to sense-data. Some of the proponents of entropic reduction of temporal asymmetry have not made this clear. Some of the work which would need to be accomplished in order to make the reduction, so understood, plausible is outlined. Then some fundamental puzzles which will afflict any such claim to a successful reduction are noted.

I

Few philosophical theses match the dramatic impact and striking illumination of Boltzmann's brilliant speculation about the reducibility of the intuitive notion of the direction of time to features of the world characterizable in terms of the theory of order and disorder summarized in the notion of entropy. Taking the progression of isolated systems toward states of highest entropy, characterized phenomenologically by the second law of thermodynamics, and given a far deeper explanation in his own theory of the statistical mechanics of irreversible processes, Boltzmann suggested that rather than viewing these theories as merely describing the asymmetric change of the world from past to future, we should find in them the very basis of our concept of the distinction between the past and future directions of time.

Building on Boltzmann's rather sketchy remarks, Reichenbach, in what many consider to be his most distinguished contribution to the philosophy of physics, elaborated for us a highly complex and subtle account of the entropic theory of time order. Yet despite Reichenbach's very impressive efforts, and the further illuminating work of others who have followed him, such as Grünbaum, Watanabe, Costa de Beauregard, and others, the claim that the very notion of temporal asymmetry reduces to that of the asymmetry of entropic processes in time remains, to say the least, controversial. To some it seems obviously true in broad outline, whatever details still need filling in. To many others the very idea of the program seems prima facie absurd.

While much remains to be done on the "physical" side of this issue, in the way of providing for us a single coherent physical account of the source of entropic asymmetry, and in the way of definitively characterizing the physical connection of this asymmetry with the other fundamental temporal asymmetries of the world, such as the outward radiation condition and the cosmic expansion, I believe that some insight into the roots of the persistent controversy can also be obtained by an attempt to become a little clearer concerning some philosophical aspects of the question about which we are yet not as clear as we might be. In particular, I think we need to be far clearer than we have been on the question of in just what sense the entropic theory is claiming that the very meaning of assertions about the direction of time are "reducible" to assertions about entropic processes. At least two fundamentally different notions of meaning reduction are available to us, and I think that confusion about just which sense the entropic theorist has in mind has served to cloud the issues in a significant way.

As a means of access into this problem I would like to make some comparisons between three different asymmetries in the world: that between the upward and downward directions of space, that between left- and right-oriented systems, and that between the past and future directions of time. I think that exploring the analogies and disanalogies between these three cases of "asymmetry" may make it clearer to us just what the entropic theorist is really claiming. While I don't believe that the insights we gain will resolve the question as to whether or not the entropic theorist is *right,* perhaps we will be a little clearer about just what both he and his opponent have a right to claim as evidence for and against the reductionist position.

II

We have a concept of the left-right distinction; and more, individual concepts of left- and right-oriented systems. We can properly identify left- (right-) handed objects; train others to do so; communicate meaningfully using the terms ("Bring me the left-handed golf clubs"), etc.

Now there may be physical phenomena described by laws which are not left-right symmetric. Current physical theory postulates that this is so, as is revealed in the familiar examples of the parity nonconservation of weak interactions; although whether this will remain the case at the level of the "most fundamental" laws remains an open question. Certainly there are many phenomena in the world which are non-mirror-image symmetric in a de facto rather than lawlike way — e.g., the preponderance of dextrose over levulose, etc. But do any of these asymmetries of the world in orientation have anything to do with what we *mean* by left and right? Is there any plausible sense in which the orientation concepts "reduce" to concepts of a not prima facie spatial orientation sort?

Most of us think not. Of course it is the case that if we wish to teach the meaning of, say, 'left' to someone, without transporting to him a particular left-oriented object, we would need to do so by means of one of the familiar features of the world lawlike or de facto associated with orientation ("Left is the orientation in which . . ."). Even then, as we know, there are the difficulties which reside in assuming that the association where he is is the same as that where we are (What if he lives in an antimatter world and CP invariance holds? What if there is more levulose than dextrose on his planet?). And, of course, if one is taking the (dubious) line that a mirror-image possible world would be the same possible world as the actual one, one would have to assume that in this mirror-image world the mirror laws and de facto correlations hold to make the Leibnizian argument (qualitative similarity implying possible world identity) go through. But none of this is sufficient to in any way back up the claim that left-handedness just is, or that 'left' just means, some relation (term) expressible in terms of not prima facie orientation concepts.

Suppose, for example, that some fairly substantial miracles occurred in this world. All of a sudden electron emission from spinning nuclei begins occurring with the dominant emission in the opposite direction from the present preferential axial direction. Would we then say that the clockwise direction had become the counterclockwise? That right-handed gloves had suddenly become left-handed? Nothing of the sort. We would indeed be astonished and look desperately for some explanation of this mirror reversal of a law. But we would, I believe, still take it that we could recognize left- and right-handed objects as before, teach the meanings of orientation terms by ostension, as before, etc.

We believe that orientation is just a basic geometric property of an oriented system. It is epistemically available to us in as direct a manner as is any geometric feature of the world. The meanings of our orientation terms are fixed for us by ostension of particular oriented objects and our facility for abstracting the right property intended by the teacher. Nothing about weak interactions is relevant here in any way. It is merely an empirical discovery of a lawlike correlation that takes places when we discover that as a matter of fact weak interactions take place in an orientation discriminating way. Were we to live in a universe in which as a matter of law, or merely as a matter of pervasive facts, all red objects were square and all square objects red, this alone would hardly be grounds for saying that redness was squareness nor that 'red' meant 'square'.

Of course our notions of handedness are complicated by reflections on the facts about dimensionality and global orientability which have been frequently pointed out. Prior to realizing the possibility that space might have a fourth spatial dimension and might be globally nonorientable we fail to notice the distinction between the partition of handed objects into two disjoint classes relative to their being constrained to the subspace in which they are imbedded and the local region of that subspace, and the global distinction which would be well defined only if the subspace exhausts the full dimensionality of space and only if that space is orientable. Under the impact of this new awareness we may want to distinguish full handedness from what I have previously called local-three-handedness.[1] And we may wish to say that the intuitive concept we had all along was the latter rather than the former. But none of this additional complication, I believe, vitiates the point here that the facts about what nonorientability features of the world are lawlike or de facto associated with handedness are irrelevant to the conceptual analysis of what we meant all along by handedness attributions.

III

Considerations like those above might lead us by analogy to make the parallel remarks about entropy increase and the future direction of time. Isn't the case just like that of weak interactions: we discover a pervasive correlation in the world, this time one whose status while not lawlike is not easily thought of as merely de facto either. But why should this in any way lead us to think that the very concept of futurity reduces in any sense to that of entropic increase?

But Wittgenstein has warned us against the deficiency diseases caused by an unbalanced diet of analogies, and we would be well advised before making a hasty judgment to look at another case which may be viewed as providing an analogy supportive of just the opposite view about time and entropy.

The suitable dietary supplement is provided for us in Boltzmann's elegant if sketchy presentation of his position in the *Lectures on Gas Theory*:

> One can think of the world as a mechanical system of an enormously large number of constituents, and of an immensely long period of time, so that the dimensions of that part containing our own "fixed stars" are

1. See L. Sklar, "Incongruous Counterparts, Intrinsic Features, and the Substantiality of Space," chap. 8 in this volume.

minute compared to the extension of the universe; and times that we call eons are likewise minute compared to such a period. Then in the universe, which is in thermal equilibrium throughout and therefore dead, there will occur here and there relatively small regions of the same size as our galaxy (we call them single worlds) which, during the relatively short time of eons, fluctuate noticably from thermal equilibrium, and indeed the state probability in such cases will be equally likely to increase or decrease. For the universe, the two directions of time are indistinguishable, *just as in space there is no up and down. However, just as at a particular place on the earth's surface we call "down" the direction toward the center of the earth, so will a living being in a particular time interval of such a single world distinguish the direction of time toward the less probable state from the opposite direction (the former toward the past, the latter toward the future).*[2]

The italics are mine.

Past and future, then, are to be viewed like up and down, and the progression toward higher entropy states like the direction of the gradient of the gravitational field (the obvious generalization of the direction of the center of the earth). It is well worth our time then to ask what we do and should say about the relationship of up and down to directions characterized in terms of gravitation and to ask for the grounds of the position we do take. We must then ask whether things are just that way with past and future and temporal directions picked out by entropic features of the world.

IV

We have a prescientific, prephilosophical understanding of the distinction between the upward and the downward direction of space. We can communicate with these concepts since they are teachable and suitable for an intersubjective language. We could teach the concept to someone either by an ostension which relies on observation of the behavior of objects (by and large they move, when unsupported, in the downward direction) or by reliance of our internal "sense" of the downward direction, using our sense of this direction to pick it out and then ostending it to another who can then identify it again by his own internal "sense" of down.

2. L. Boltzmann, *Lectures on Gas Theory 1896 – 1898,* trans. S. Brush (Berkeley: University of California Press, 1964), 446 – 447.

Naively we view it as a global notion, in the sense that parallel transport of a downward pointing vector keeps it downward pointing.

But then we discover gravitation. We come to understand that it is the local direction of the gradient of the gravitational field (on the surface of the earth, the local direction of the center of the earth) which "picks out" at any point the downward direction. "Picks out," though, in a deep sense. It isn't just that the local gradient happens to point down, nor even that the local direction of the gravitational gradient points downward as a matter of lawlike necessity. Rather it is the reference to the local behavior of the gradient of gravity which offers a full and complete account of all those phenomena which we initially used to determine what we meant by the downward direction in the first place.

Understanding gravity, we understand why, in general, objects fall downward. We even know, understanding gravity and a few other things as well, why helium balloons, flames, etc., don't. A complete, coherent, and total explanation of all the phenomena we associated with the notion of 'down', associated in the sense of used to fix the very meaning, or at least reference, of 'down' is provided for us by the theory of gravitation.

We even know (although only vaguely to be sure) why it is that we can pick out the downward direction by an "internal" sensation; why it is that we can know which way is down without ever observing an external falling object. The explanation has to do with the forces, once again gravitationally explained, exerted on the fluid of the inner ear. A demonstration of this and a full account is presumably a matter of some complexity, but we can rely on inference from the behavior of simpler creatures. There are fish with sacks in their bodies with sand in the sack. Remove the sand and replace it with iron fillings. Place a magnet over the fish tank and the fish swim upside down. Surely it is something like that with us. In any case we don't doubt but that the ultimate physiological account of our inner apperception of down will refer ultimately to the effect of gravitational forces on some appropriate component of a bodily organ.

How should we describe appropriately the relationship between 'down' and 'the direction of the gradient of the gravitational field'? I am looking here, not for the ideal description in our ideally worked out semantico-metaphysics, but rather for the sorts of things we are, initially, intuitively inclined to say.

It wouldn't be strong enough to say that the downward direction is the direction of the gravitational gradient, for that would be true were it merely a happenstance that down and the direction of the gradient coincided. We feel, rather, that the downward direction is "constituted" (whatever that means) by the direction of the gradient. Perhaps the correct locution is: Down (the downward relation itself) *is* (is identical to) the relation between points constituted by one's being deeper in the gravitational potential than the other. We *identify* the relation of *a*'s being downward with respect to *b* with *a*'s having a lower gravitational potential than *b*. (It is more complicated than that, of course, since *b* could be very remote from *a*, in which case we wouldn't talk that way if there were, for example, intervening regions of higher potential, but I am deliberately going to oversimplify grossly here.) Put this way the "reduction" of the up-down to the gravitational relationship bears close analogy with substantival identifications as a means of theoretical reduction (water is H_2O, light waves are electromagnetic waves). But it is a property (relation) identification rather than one of substances.

I think that some would want to go further, arguing that the reduction established is sufficient to allow us to say that the very *meaning* of 'down' is given by the appropriate characterization of a relation in terms of the gravitational gradient. Now meaning is a notion as yet sufficiently unconstrained by a real theory as to allow us, with some plausibility, to say any one of a number of different things. Emphasizing the connection of meaning with criteria of applicability (verification procedures, operational definitions, etc.) we would be inclined to say that although 'down' doesn't (or at least didn't) *mean* 'the direction of the gravitational gradient,' what was empirically discovered was that the downward relation was the relation gravitationally described. From this point of view there is a change of meaning which has taken place when scientists, now fully aware of the gravitational account of the up-down phenomena, begin to simply use 'down' to mean the local direction of the gradient of the gravitational field.

Emphasizing, on the other hand, the association of meaning with reference, in the manner of some recent semantic claims about proper names and natural kind terms, we might, instead, be inclined to say things like: "'Down' meant, all along, the local direction of the gravitational gradient." Of course it is still a discovery on our part that gravity plays the explanatory role it does. From this point of view we might even be tempted to say that prior to the full understanding of the gravitational explanation of up-down phenomena, people simply didn't understand what they meant by 'down'. And one will, of course, now begin to claim that it is a necessary truth that the downward direction is that of the gravitational gradient, allowing into one's scheme the now familiar necessary a posteriori propositions which result from such a "referentialist" semantics.[3] I do not wish to discuss any of the arguments for or against such a view of meaning here, but only to emphasize, once again, that insofar as a reduction of the up-down relationship to one characterized in terms of the gravitational gradient is plausible at all, it is a reduction which bears very striking analogies to the reductions by means of substantival identification so familiar to us in other cases of intertheoretic reduction. It bears an even closer resemblance to such property identifications as the familiar (if abused) example of philosophers "Temperature just is (is identical to) mean kinetic energy of molecules." Oversimplified as that claim may be, the essence of what it is getting at is surely correct. "Down just is (is identical to) the direction of the gravitational gradient" seems a claim of the same order and, if anything, probably in need of fewer qualifications and reservations than are required in the thermodynamics-to-statistical mechanics case.

It is important to emphasize at this point the kind of reduction which the one in question certainly is not. Perhaps no one would ever, in this context, make the kind of mistake I am warning against here, but I believe that in the context of the problem of the direction of time just such a confusion of kinds of reduction has played some role in muddying the waters. In saying that up-down reduced to the gravitational gradient relation we are not making a claim based upon a notion of priority of epistemic access. Such a claim is familiar to us in the claims that material object statements "reduce" to sense-datum statements, spatiotemporal metric statements "reduce" to statements about the local congruence of material measuring instruments, etc. In the present case, unlike the ones

---

3. See S. Kripke, "Naming and Necessity," in D. Davidson and G. Harmon, eds., *Semantics of Natural Language* (Dordrecht: Reidel, 1972), 253–355.

just cited, there is no claim that our epistemic access to the up-down relationship is mediated through any sort of "direct awareness" of the gravitational relationship; nor that some kind of hierarchy of epistemic immediacy tells us that up-down statements are, while initially thought of as inferred from gravitational statements, actually translatable into logical complexes of the gravitational type statements. Instead the claim is just that the up-down relationship is found, by empirical research, to be identical with a more fundamental relationship characterizable in terms of the gravitational field. Down is the direction of the gravitational gradient as water is H_2O, light electromagnetic radiation and temperature mean kinetic energy. Not as tables are logical constructs out of sense data nor as nonlocal congruences are logical constructs out of spatiotemporally transported rods and clocks.

It will be useful at this point to say a little about one further aspect of the reduction of up-down to that of the gravitational gradient. Prior to understanding the gravitational nature of down we intuitively viewed the downward direction in a global way: at every point the downward direction was parallel to the downward direction at every other point. (Of course this description of the situation is something of a travesty of the way in which the conceptual change occurred slowly over time. Aristarchus was well aware of the spherical nature of the earth and probably quite cognizant of the fact that down at Thebes was not parallel to down at Athens.) Recognizing the gravitational nature of the up-down relation we now realize clearly that what is down for us will most certainly not be parallel to what is down for someone at a different point on the earth's surface. We even understand that at some points of space there really won't be any downward direction at all.

Of course many ways are open to us to describe this. We can, if we wish take 'down' to mean the direction of the gravitational gradient at the place we are located, identifying the downward direction elsewhere as the direction at that point parallel to our down. From that point of view Australians do, indeed, live their lives out upside down. We might, to eliminate confusion, introduce a nondenumerable infinity of subscripted "downs," 'down$_p$' referring to the downward direction at the point referred to by the subscript. Then Australians live upside down$_{USA}$ but, of course, right-side-up$_{AUST}$. Alternatively, and more elegantly, we can simply take 'down' as having an unequivocal meaning but as functioning in the manner of a token-reflexive, at least to the extent that:

(1) what is referred to as the downward direction by a speaker at one place is the direction of the gravitational gradient at that place;

(2) what is referred to as downward by a speaker at another place is the direction of the gravitational gradient at that place;

(3) and there is no reason whatever for thinking a priori that the referents of the two utterances of 'down' will be the same.

From this point of view, there is a clear sense in which the *sense* of 'down' is the same for all speakers at all places.

V

I think it is clear that the entropic theory of temporal direction, if it is to be plausible at all, should be viewed as a "scientific" reduction motivated by an empirical discovery of a property (relation) identification, and not as an instance of the "philosophical" reductions motivated in terms of a critical analysis of the modes of epistemic access to the world available to us. Perhaps this is obvious to many. But it hasn't always been obvious to me, and at least some others have been misled. The following quote, for example, is, perhaps, indicative of this confusion of modes of reduction. I think it is appropriate here, even though it refers to a causal theory of the direction of time, since, after all, the theory has been for many years an entropic theory of temporal direction rather than a causal theory.

It is sometimes suggested that the direction of time and causation are linked because the direction of time is itself to be analyzed in terms of causation. But, at least as *conceptual analysis,* this must be wrong. We can think of events' succeeding one another in time even if there are no causal links between any of them, let alone between the members of each pair of which one is earlier than the other. Moreover, *our concept of the direction of time is based on a pretty simple, immediate, experience of one event's following straight after another,* or of a process going on — say of something's moving — with a later phase following an earlier one. It might be, of course, that our having such experiences is somehow dependent upon causally asymmetrical processes going on inside us — we might have internally causally controlled unconscious temporal direction indicators — but even if this were so it would not mean that our concept of time direction was analyzable into that of causal direction. *Our experience of earlier and later, on which our concept of time direction is based, itself remains primitive,* even if it has some unknown causal source.[4]

But if the entropic theorist has in mind reduction of the "scientific" kind, then nothing in the way of immediate, simple experience of earlier and later events, or ongoing processes, nor any reference to an ability to imagine (think of) events being temporally ordered without being entropically related will refute the claim, meant in this sense, that the later-than relation is (is identical to) some relation characterized in terms of entropy, nor even that, in the senses of meaning we noted above, in some sense 'later than' *means* 'bears some appropriate entropically characterisable relation to'.

Since the two notions of concept reduction I have been discriminating are easily confused in general, it isn't too much of a surprise that we have not always been clear which sense of reduction is intended by the entropic theory of temporal direction. But I think that some of the very arguments used by entropic theorists have tended to ingrain the confusion. For example, entropic theorists frequently ask us to consider how we would distinguish a film of events run in the proper order from the film run in reverse order, pointing out to us that the discrimination can only be done (or, rather, so it is claimed) when entropic features of the world are present, and that it is by means of the expected dissipation of order into disorder that we make the judgment about whether the film is being run in the correct direction. If this is meant only to show us that the entropic features of the world are, at least, the most prominent which are asymmetric in time order and, hence, the prime candidates for a reduction of the scientific kind, then it is harmless. But it is easy to slip from this argument into the dubious claim that we judge time order events in the actual world by inference from

4. J. Mackie, "Causal Asymmetry in Concept and Reality," unpublished paper presented at the 1977 Oberlin Colloquium in Philosophy, 1. The italics are mine. For other expressions of skepticism about the entropic theory, see J. Earman, "An Attempt to Add a Little Direction to 'The Problem of the Direction of Time'," *Philosophy of Science* 41 (1974): 15–47; and L. Sklar, *Space, Time, and Spacetime* (Berkeley: University of California Press, 1974), 404–411. Most important, see A. Eddington, *The Nature of the Physical World* (Cambridge: Cambridge University Press, 1928), chap. 5, "Becoming."

apprehension of ordering of states in respect to entropy. As Mackie and others have pointed out, this is indeed dubious. But the dubiousness of that latter claim is an argument only against the "philosophical" theory of the reduction of time order to entropy. In no way would it vitiate a reductive claim of the "scientific" kind.

Again consider Reichenbach's transition from a causal to an entropic direction. If what is being said there is that the only relevant causal notion is causal connectibility, that this is temporally symmetric, and hence not a suitable candidate for a reduction basis for the relation of temporal order, then it is a point relevant to reductions of the identificatory kind. But it is easy to read the argument as saying that the causal theory won't do because we must be able to empirically *determine* which of a causally connected pair of events is cause and which effect in order to make the reduction go through, and that this determination requires first *knowing* the time order of events, and that this makes causation unsuitable as a reduction basis for temporal direction because it lacks the necessary epistemic independence and primacy. But this latter argument, once again, suggests that it is the epistemically motivated kind of reduction which the theorist has in mind. If he then offers an entropic theory as the substitute for the causal, one is misled into thinking that the theory too is an attempted reduction of the "philosophical" sort.

There is also the fact that Reichenbach presents the entropic theory as part of a general reductivist account of spacetime. Entropy is to fix one last part of spacetime structure, the past-future distinction, after the rest of spacetime, in particular its topology, including its temporal topology, has already been "reduced" to non-prima-facie spatiotemporal notions. In particular, the spacetime topology is supposed to be reduced to the *causal* structure of the world.

Now I think that a kind of "scientific" identificatory reduction of spacetime topology to causal order could be argued for. For example the recent suggestions of reducing spacetime structure to some kind of algebraic relationship among quantum events might be viewed as a reductionist move of this kind. But I think that the causal theory of spacetime topology which Reichenbach offers is, rather, motivated by, and formulated in terms more appropriate to, an epistemically generated type of "philosophical" reduction. If this is correct we can see why one would easily be misled into thinking that the entropic account of time direction was also supposed to be a reduction of this latter kind.[5]

VI

But if the entropic theory of the direction of time is supposed to be a scientific reduction we must ask whether or not it is successful. Is the connection of entropy with time order, like that of asymmetric weak interaction processes with left and right, merely a correlation (lawlike or de facto), or is the case rather like that of gravitation and up and down, where we feel it is at least appropriate to say that the up-down relation is identical to the gravitationally characterized relation, and where we are even tempted (at least on some theories of meaning) to say that 'down' means 'in the direction of the gradient of the gravitational field'?

That question I hardly intend to try and answer here. What is needed is a full-fledged attempt to try and account for all the processes we normally (prescientifically) take to mark out the direction of time, including our internal "direct" sense of temporal order, in terms of a single, unified account which invokes the relation of difference in entropy and accounts for all these phenomena in terms of an identification of the time order relationship with some relationship among events characterizable (at least in part as we shall see) in entropic terms, and which does not invoke time order itself as a primitive in the characterization. Despite Reichenbach's heroic efforts in this direction, I think we can all agree that such an account is not yet available to us.[6] But, of course, Reichenbach's efforts, from this point of view of the nature of the entropic theory, are at least efforts in the right direction. We must explain, entropically, why causes precede their effects (at least usually); why we have records of the past and not the future; why we know and believe so much more about the past than the future, and believe and know about them in such very different ways; why we feel we can change the future but not the past; why we have such a different emotional attitude toward the future than we do to the past ("Thank God that's over!"); why we take the past to have deter-

5. On the causal theory of spacetime topology as an instance of philosophical reduction, see L. Sklar, "What Might Be Right about the Causal Theory of Time," chap. 9 in this volume.
6. See H. Reichenbach, *The Direction of Time* (Berkeley: University of California Press, 1956).

minate reality and the future to exist, if at all, merely as "pure potentiality"; and, finally, why we have direct, immediate, noninferential knowledge of the time order of events (internal and external) with which we are directly acquainted (in Russell's sense).

While many of Reichenbach's arguments in these directions are brilliantly imaginative and suggestive, I do not believe that I will be taken to be disrespectful if I assert here that they are, to many of us, far from conclusive. They serve as brilliant suggestions toward a theory, but the theory we will ultimately be given by the entropist as highly confirmative of his reductive claim is still in the future. Here I wish only to make a few rather general remarks about the entropic program, some of the difficulties it faces, and why at least some suggested objections to it are not really devastating to its aim.

(1) At least part of the problem in trying to establish the entropic theory is the rather vague grasp we have on many of the notions to be accounted for entropically in the reduction. Compare asking: "What is a causal relation?" "What is a record or trace of an event?" etc., with asking "What is a falling object?" In the gravitational theory of down at least we have, prior to the reduction, a pretty good idea of what it is that the gravitational theory must account for. In the entropic theory of time direction we don't have a very clear idea at all. Of course, it may very well be claimed by the entropic theorist that it is only in the context of the reduction that our ideas of what it is that must be accounted for will become clear. I think that Reichenbach has this in mind. For example, only when we understand the role played by entropic features of the world in our prescientific conceptual scheme will we really begin to understand our preanalytically felt, but very poorly understood, intuition that causal efficacy proceeds from past to present and thence to future.

(2) I have deliberately avoided any attempt at saying exactly what relationship among events, characterized entropically, is the one to which the 'later than' relationship is to be identified. It is clear that this identification will be one of some subtlety. In the case of the up-down relationship the identification is fairly simpleminded. If b is downward from a then there is a gravitational potential difference between them determined by the value of the potential at the two points. That plus some facts about the gravita-

tional potential at intermediate points is enough to fix the appropriate "gravitational" relationship in our reduction basis.

But the temporal asymmetry case is trickier. First of all there is the fact that a later state of even an isolated system can very well be one of lower entropy than an earlier state. We must take account of the fact that the association of entropy order with time order is supposed to be only statistical. Second, there is the fact that we take the time ordering relationship to be more pervasive than that of entropic order, in the sense that there can be a later-than relationship holding between events where no obvious characterization of the events as states of affairs of a system with different entropy is at all possible.

Now one solution to this would be to postulate the existence of a "time potential" with a gradient in the timelike direction, making the time order case look far more like the up-down case than even I have maintained it to be. This is Weingard's suggestion in a 1977 article.[7] I think this is the wrong way to go. I don't deny that in some possible world that is how things could turn out to be, with the existence of such a time-ordering vector field as a "real physical field" whose existence is ultimately explanatory of the familiar asymmetries of the world in time. It is just that we have no reason whatever to believe that in this world there is any such field. The usual statistical mechanical explanations of the asymmetric behavior of systems in time invoke no such fundamental field. Granted we frequently do find the statistical mechanical explanations unsatisfying, and many have the feeling that at the present state of our understanding some matters of fundamental importance have yet to be uncovered. But few physicists would presently accept as a plausible explanation the existence of such a fundamental time-ordering field as the underlying "missing link" in attempting to offer a full explanation of the asymmetry of the world in time. If the theory of time direction is supposed to be a scientifically established identificatory reduction of the later-than relation to some other more fundamental relation, then it must be established by the real science of the world as it actually is. A possible reduction which would be satisfactory in some possible, but nonactual world, is of no help to us.

Nor need we invoke such a pseudofield in order to have an

7. R. Weingard, "Space-Time and the Direction of Time," *Nous* 11 (1977): 119–132.

adequate account. One direction in which to move is again available to us in Reichenbach. The whole entropic theory presupposes an underlying theory of time order — the full topology of time (or spacetime) — with the only intuitive feature removed being that of the past-future asymmetry. Of nearby pairs of pairs of events (nearby to avoid the possibility of non-temporally-orientable spacetimes) we can ask if d is in the same time direction from c as b is from a. If we can then establish the "laterness" of, say, b to a on the basis of entropic considerations, we can "project" this time order onto the $c-d$ pair, taking d as later than c, even if none of the relevant entropic considerations appear in the $c-d$ case. Actually, of course, the detailed theory might be much more complicated than that, making reference, possibly, to multitudes of systems and where entropies can be assigned to temporally distinct states of isolated systems, to entropic difference "parallel" for the overwhelming majority of them. Then the past-future time direction is taken as being fixed by this majority of systems, lower entropy states being earlier than higher entropy states, and thence "projected" by local comparability of time order to all pairs of temporally related events. For our present purposes the details are inessential.

(3) What I have suggested above suggests an approach to the entropic account which offers a "definition" for time direction in terms of the entropic behavior of branch systems, in Reichenbach's terminology.[8] I might say something here about the relevance of the branch system notion to the overall account. With what entropic feature of the world do we wish to identify the time direction? Not if Boltzmann's overall approach is correct, the entropic relationship between states of the universe as a whole (assuming that such a notion as entropy for the universe as a whole is well defined, a powerful and dubious assumption). More plausible would be an identification of the later-than relation at a placetime with the appropriate entropic relation among the states of the "single world" during the "eon" containing the placetime. This, indeed, might be the right direction for the entropic theorist to go, rather than that outlined above.

Why need we invoke the branch systems of Reichenbach? If we were holding to an epistemically motivated "philosophical" reduc-

8. See Reichenbach, *Direction of Time,* 118–143.

tion, the answer would be obvious. We have, certainly, no epistemic access of a direct sort to the total entropy even of our "single
world." But, perhaps, we do to local temporarily isolated systems.
So we observe them, and the entropic relationships among their
states, and "infer" time order from these relationships. The reduction then consists in replacing this "inference" with a "coordinative definition" in the familiar way.

But I have been maintaining that it is not this kind of reduction
which the entropic theorist is really after. What then is the role of
branch systems and their entropically characterizable states? I
think it is that in our explanations of the various phenomena characteristic of the asymmetry of the world in time intuitively associated with the time order of the world in our prescientific picture,
the branch systems and their states will have to be invoked. Even if
ultimately in the explanation of these asymmetries we refer to the
entropic behavior of our "single world" during its present "eon,"
the explanation will invoke at an intermediate stage some account
of how this entropic asymmetry gives rise to the entropic asymmetries of the branch systems, and will then use these "small" entropic
asymmetries to account for the familiar asymmetries of causation,
knowledge, traces, etc., and to account for our immediate internal
sense of the time order of our own experiences.

Whether the entropic theorist will then want to identify the
later-than relationship with a relationship entropically characterized among the total states of the "single world" or, instead, with
some complex "majority rule" relationship among states of sets of
branch systems I do not know. I think we would need more detail
about the nature of the entropic theory to decide this. Or, perhaps,
he has a choice and there is an element of arbitrariness in the
identification he asserts.

(4) We saw that in the gravitational reduction of the concept of
up and down to that of the direction of the gravitational gradient, it
was no argument against the account at all that at different places in
space the downward direction could vary. The same holds true with
the entropic theory of time direction. Whether or not Boltzmann is
right that there are at a given time "single worlds" with their time
orders oppositely directed, or a single "single world" which at different "eons" has its time order in the reverse direction, this is
certainly a possible state of affairs on the entropic account. And

nothing about this state of affairs makes the entropic theory in any way less plausible.

Once again we have a choice of at least two ways of describing the situation. We can take the future direction of time as fixed by the entropic relations among states of our "single world" in our "eon" and speak of entropy as going the "wrong way" in time in the counter directed worlds. Or, less parochially, we could take the past-future relationship to be quasi-indexical, letting 'future' refer to that direction in time at a spacetime point which is the direction bearing the appropriate entropic feature in that "single world" during that "eon." None of this is incompatible with the earlier remarks that the entropic theorist might wish to use *local* comparability of time order to project the past-future relationship from some system in his "single world" to others.[9]

VII

At this point my already very sketchy and somewhat vague essay is going to become even less the presentation of a polished, finished account. For I am here going to suggest that, at a new level, some of the standard objections to the entropic account may reappear, even if that account is interpreted in its most plausible form as an account of a "scientific" identificatory reduction.

At some point the reductive programs of the naturalistic sort which proceed by identificatory reductions of substances and properties to those more "scientifically" fundamental, and the reductive programs of the philosophical sort which proceed by "conceptual analysis" of propositions in terms of a critical examination of the total class of propositions which could serve as epistemic warrant for them, must be reconciled. One could, of course, reject the latter kind of reductionism altogether as spurious, but I don't think that we can do this without at the same time rejecting some of the deepest and most well accepted portions of our recent scientific progress; for, I would allege, much of the transition from space and time to relativistic spacetime proceeds by just such an epistemically motivated "reductionist" critique. I won't argue this here, but only try to show how one aspect of "scientific" reductions introduces, in

9. The arguments here are in reply to an argument of Earman's, especially to his invocation of what he calls the Principle of Precedence. See his "An Attempt to Add a Little Direction," 21–23.

the particular case of the entropic account of the direction of time, some special difficulties closely related to the problem of working together these two kinds of reductive analysis.

A familiar concomitant of identificatory reductions is the "secondarizing" of properties. Tables are arrays of atoms. But what about the "immediately sensed" properties of macroscopic tables? Are they properties of arrays of atoms? Arrays of atoms are, in some sense, discontinuous; but what of the sensed, continuous color patch that a table presents to my awareness? One solution (maybe not the only and maybe not the best) is to remove properties from the table (except for leaving a residuum of them as powers or dispositions) and reclassify them as secondary qualities of sense-data or, perhaps, of the sensing perceiver (who is appeared to reddishly, etc.)

Temperature is mean kinetic energy of molecules. But what of the felt quality we first used to discriminate hotter from colder objects? Easy, make it a secondary quality "in the mind" of the perceiver. Whenever we propose an identificatory reduction of some entity or property, initially identified by us by a "direct apprehension," to some other entities or properties in the world, there is at least the temptation to strip off from the object the original identifying feature and place it "in the mind" as a secondary quality related to the reducing property in the world only as the causal effect of that property's acting, by means of the sensory apparatus, on the "mind." I'm not saying that this is the only direction in which to go nor that it is the right one; only that it is a persistent, common move and one intuitively hard to resist.[10]

Now we take the later-than relation to be a relation in the world characterizable in entropic terms. But what of the "pretty simple, immediate, experience of one event's following straight after another"? Our temptation is, I think, once again to dissociate the immediately sensed, directly apprehended, "later-than-ness" of events from the time order of events in the world, making it into a feature only of events "in the mind."

But now we see why many who would easily accept the claim that tables are, in fact, arrays of atoms, and that temperature is, in fact, mean kinetic energy of molecules, will balk at the claim that later-than-ness is an entropically characterizable relation among

10. See L. Sklar, "Types of Inter-Theoretic Reduction," *British Journal for the Philosophy of Science* 18 (1967): 122–123.

events in the world. We feel that time order is something that holds of events in the world and the events of inner experience as well. Since Kant we have been familiar with the claim that space is the manifold of experience of outer objects and time of both inner and outer awareness. But it is the same time which relates outer events and which relates events "in the mind." And if outer events are later than one another, are they not later than one another in exactly the same sense that inner experiences occur in the asymmetric order of time? And if I directly experience this order among events in my inner mental life, mustn't I identify that relation with the real later-than relation among events in the world? If these events in the world are also related by some relation entropically characterizable, mustn't that be viewed as an empirically established correlation with time order then? And isn't it then true that there is no more plausibility in identifying the later-than relation with the entropic relation than there is in identifying left-handedness with some feature of an object characterized in terms of the behavior of weak interactions?

Notice the difference here from the gravitational case. Our inner experiences are not, really, up and down from one another. No harm then in disassociating our inner experience of down from the real down-ness relation in the world and then identifying the latter with a relation characterized in terms of the gradient of the gravitational potential. But inner events are *really* later and earlier than one another and our "pretty simple, immediate experience" of this relation cannot with impunity be detached as merely a causally induced secondary quality not properly thought of as a direct experience of the real afterness relation which exists in the world only as an entropically characterizable relation.

The following quote from Eddington suggests that it is something like this argument which is at the root of many of the strongly felt but not very well expressed objections to the plausibility of the entropic theory. It is important to note that this quote is from one of the earliest expositors of the entropic theory of time direction as I have described it.

> In any attempt to bridge the domains of experience belonging to the spiritual and physical sides of our nature, Time occupies the key position. I have already referred to its dual entry into our consciousness — through the sense organs which relate it to the other entities of the physical world, and directly through a kind of private door into the

mind. . . . Whilst the physicist would generally say that the matter of this familiar table is *really* a curvature of space, and its color is really electromagnetic wavelength, I do not think he would say that the familiar moving on of time is *really* an entropy-gradient. . . . Our trouble is that we have to associate two things, both of which we more or less understand, and, so as we understand them, they are utterly different. It is absurd to pretend that we are in ignorance of the nature of organization in the external world in the same way that we are ignorant of the intrinsic nature of potential. It is absurd to pretend that we have no justifiable conception of "becoming" in the external world. That dynamic quality — that significance which makes a development from future to past farcical — has to do much more than pull a trigger of a nerve. It is so welded into our consciousness that a moving on of time is a condition of consciousness. We have direct insight into "becoming" which sweeps aside all symbolic knowledge as on an inferior plane. If I grasp the notion of existence because I myself exist, I grasp the notion of becoming because I myself become. It is the innermost Ego of all which *is* and *becomes*.[11]

I don't pretend to understand all that Eddington is saying here, nor to be able to give a really coherent version of my own arguments above. I do think, however, that it is very clear that our ultimate view of the world will require a subtle and careful weaving together of the naturalistic reduction of science which proceeds by theoretical identification with the conceptual reduction of philosophy which proceeds by epistemic analysis. Until we have such a systematic overall account I think that the ultimate status of an entropic theory of time order will be in doubt.

11. Eddington, *The Nature of the Physical World*, 91–97.

Index

Acceleration, 199, 202, 210, 247

Aether theory: and observational/ nonobservational distinction, 175, 180, 181, 184–185, 187, 209, 210, 279; and special relativity, 5, 12, 167, 173, 174, 178, 179

Alexandroff topology: and causality, 254, 274, 295; and manifold topology, 140–141, 142, 145n; and Robbian formalization, 81, 114, 116; and Woodhouse formalization, 143, 144. *See also* Topology; Zeeman topology

Analogy: semantic, 215–233; and entropy, 309

Analytic/synthetic distinction, 95, 96, 118–119, 120, 272, 274, 277, 287

Apriorism: and causal theories, 272, 283, 284; and meaning of terms, 219; and naturalism, 170, 177, 178, 179; and theory choice, 100, 104, 175

Aristarchus, 314

Asymmetry: entropic, 1, 3, 281, 282, 283, 305, 307; and general relativity, 13, 14; temporal, 17, 18, 289, 290, 300, 301, 305–307, 315–322, 325. *See also* Handedness

Bayesian theories, 152, 162

Believability, 63–64, 158–159, 172, 224

Bergson, H., 91, 123

Berkeley, G., 229

Bohr, N., 230

Boltzmann, L., 288, 306, 309, 321, 322

Bondi, H., 192

Cartan, E., 196

Cartesian theories, 59, 60, 176, 187, 286. *See also* Descartes, R.

Causal connectibility: and congruence, 137; and conventionality, 134, 140, 297; and defined/primitive terms, 109, 110, 111, 112, 113; in EPS formalization, 132, 133; in general relativity, 130, 131; and reality, 294, 298; in Robbian formalization, 85, 86, 88–90, 93–95, 97, 99, 102, 104–108; in special relativity, 108; and topology, 114, 140–145, 295, 296

Causality: and direction of time, 315, 317, 319; in Minkowski spacetime, 79, 81, 82, 83, 84, 142, 268; and spatiotemporal order, 15–19, 73, 74, 75; in theories of the metric, 115–126, 129–140; in theories of spacetime, 268–288; and topology, 140–145, 249–267

Clarke, S., 9

Confirmation, theories of, 8; and conservatism, 41n7; and observational/nonobservational distinction, 178; and theoretical equivalence, 52, 53, 62, 63, 65, 66, 70–72; and unborn hypotheses, 152

303; and theoretical equivalence, 90; and time, 300–301, 302, 304

Watanabe, S., 306
Weingard, R., 320
Weyl, H., 81, 83, 130, 136
Wheeler, J., 130, 155
Winnie, J., 74, 80–81, 115, 126

Wittgenstein, L.: on analogies, 309; on empiricism, 169; on meaning, 215, 221, 223, 224–225, 229
Woodhouse, N., 124, 142–143, 144

Zeeman, E., 82, 83, 259
Zeeman topology, 114, 122, 143–144, $276n$

Designer: Barbara Llewellyn
Compositor: Progressive Typographers
Text: 11/13 Times Roman
Display: Times Roman
Printer: Thomson-Shore, Inc.
Binder: John H. Dekker & Sons